A Vedic Life

The Practical Philosophy of Nature

'A Vedic Life' is a presentation of some of the important features of the Vedic approach to knowledge. It is not restricted to any particular person or book, though it is closely associated with the Vedas as they represent its open and inclusive approach to knowledge.

A Vedic Life

The Practical Philosophy of Nature

PAVAN KANWAR

New Age Books

A VEDIC LIFE

ISBN: 978-81-7822-373-5

First Edition: Delhi, 2011

pavan.kanwar@yahoo.com

Published by
NEW AGE BOOKS
A-44 Naraina Industrial Area Phase-I
New Delhi-110 028 (INDIA)
Email: nab@vsnl.in
Website: www.newagebooksindia.com

Printed in India
at Shri Jainendra Press
A-45, Naraina Industrial Area, Phase-I,
New Delhi-110 028

To Ganapati

May we feed him with delightful thoughts,
So that he may grow within us with a
powerful head and strong body,
To support us through the maze of our actions.

To the River Sarasvati

May she be fed by the melting snows of the
peaks of our consciousness,
So that her delightful waters may flow to the
planes of our intellect,
And nurture inspirational thought and speech.

Dedication

Humbly dedicated to the vision of the Practical Vedanta
of Swami Rama Tirtha,
Which we lost in the river of inspiration.

Thanks

To all the teachers and guides
every where and at all times.
In particular to Dr. David Frawley
for his wisdom over the years.

Contents

Foreword

Many of India's great spiritual teachings have gained significant recognition and followings worldwide over the past two centuries. This includes Yoga, Vedanta, Tantra, Ayurveda, Vedic astrology and the Vedas themselves. It involves such practices as asana, pranayama, mantra, meditation, ritual, and various healing practices. There is no country and no major city where these teachings are not represented to some degree today and the interest appears to be growing continually.

Yet these great teachings of India - which are largely those from the Vedic tradition - though sharing a common heritage and having many historical interconnections, are generally studied in isolation, with practitioners and teachers of one approach not usually informed about all the others. Sometimes these different Vedic teachings are looked at not as related facets of a single greater teaching, but as unique, unconnected or unrelated, or even as opposed to one another. Scholars, particularly those who are not practitioners in such traditions, emphasize differences that are largely semantical in nature. They often promote their own social, political and intellectual interpretations and miss the deeper flow of spiritual knowledge and inner practices behind these teachings as adapted to different temperaments, times and circumstances. No doubt differences do exist between such Vedic and Indic teachings but these are relatively minor, constituting different flavors or rasas as it were, not an entirely different approach, conflicting values or a different culture. Academics tend to divide these Indic traditions into separate lines or even place their origins outside of India or outside the Vedic tradition, failing to see how connected all these teachings and practices have always been, particularly at the level of inner experience.

The result is that students in these Vedic and yogic teachings seldom understand how their particular discipline is connected to the others, even though a clear common ground is present in the traditional literature, if not in their own lineages. More importantly, such a one-sided learning also leaves students without the complete tools to deal with life as a whole, not allowing them to understand how such Vedic disciplines as Ayurveda or Vedic astrology, for example, can help with their meditation or mantra practices. They fail to see that all these teachings do serve different purposes as dealing with the different aspects of our lives, like the various branches of a single tree.

Such misunderstandings are particularly true of Yoga, which is the most popular of these Vedic and Hindu traditions in the world today. Few modern Yoga teachers, much less students know, for example, how deeply the classical Raja Yoga of Patanjali and the *Yoga Sutras* connects to the *Vedas*, Vedanta, Samkhya, Ayurveda and even Tantra. Though the *Yoga Sutras* are the main text of the Yoga Darshana which is one of the Shad Darshanas or the six systems of Vedic philosophy that accept the authority of the Vedas; though the *Yoga Sutras* are a Smriti text in the different classification of Hindu literature accepting the Vedas as their Shruti or revelation, many interpreters of the *Yoga Sutras* fail to connect Yoga with Vedic thought, Hindu/ Sanatana Dharma, or the greater Yoga tradition of which the *Yoga Sutras* are just one important but not a unique text.

Today, however, this divisive trend is slowly beginning to change, particularly as the Vedic sciences are gaining greater regard. Many in the Yoga field are being exposed to the Vedic philosophies, arts and culture. They are beginning to use Ayurvedic medicine, Vedic astrology, Vedic mantras like the Gayatri, deities, yajnas and pujas. The current generation of gurus from India is also bringing these related Vedic teachings more commonly into their practices, with Ayurvedic medicine or Vedic rituals available at their ashrams and centers.

In this context it is very refreshing to see a book that in simple, clear and comprehensive terms shows us how each of these Vedic and yogic teachings are part of a bigger picture and work best when used together for their complementary advantages. The book, as its title indicates, aims at a Vedic life. It looks at the Vedic arts and sciences in an integral manner relative to the whole of human life and the greater life of the universe, the Cosmic Purusha. The title of the book reminds us that such Vedic teachings are not merely an academic concern or of historical and cultural interest only. They are not limited to monks and sadhus or addressed only to renunciates. Studying them is not a matter of merely examining ancient texts, obscure references and complex systems of terminology. These Vedic teachings are part of a greater science of life and consciousness that is relevant to every person, every day life, and can be explained in such a way that they can be understood and utilized by every sincere seeker.

Pavan Kanwar is a rare author who has actually studied these various Vedic teachings in a traditional setting and practiced them for a number of years. He can speak both with the authority of examination and application of not only the texts but their practices. Kanwar is part of a new generation of scholars and teachers who are not content to look at the Vedic teachings from the outside as a relic from the past, but who have immersed themselves and developed both their outer and inner lives according to an application of the Vedic teachings in their own unique circumstances in life. The view of such 'practitioner scholars' provides a new, deeper and more living dimension to traditions and texts, which though rooted in ancient times, reflect values and principles that are applicable to every generation.

Kanwar does not limit himself to the philosophical and rational side of the Vedic teachings, as modern scholars have tended to do. He also shows the relevance of the Devas, the great deities that are not simply a primitive polytheism but a sophisticated science of how the Divine consciousness takes various names, forms and functions on different levels of the psyche and the cosmos. He links Vedic sciences, particularly mantra, ritual and deities, to Tantric sciences, which are usually looked at only according to their sensational side. He brings in Ayurvedic medicine and Vedic astrology and does not deal with the spiritual and philosophical teachings only apart from their life application and cosmology.

For those wanting a good introduction to the greater Yoga/ Vedic tradition and its most important applications, this is a very helpful book and one that communicates to both the East and the West. It condenses in a short volume a comprehensive study of many teachings and traditions. It is not simply a modern synthesis but reflects the ancient synthesis at the core of these apparently diverse teachings. In its pages, the Vedic teachings are alive and relevant, growing and expanding anew, not simply specimens for a museum or an old library.

Kanwar shows that the inspiration and insight to a greater connection with the unbounded conscious universe pervades all these Vedic teachings, and can still be accessed to bring their power into our own life experience, wherever we may reside in the world. The inner Sarasvati River still flows on and continues its pull, drawing us into the infinite ocean of awareness. Such books help us connect with her vibrant currents. These books reflect the living teachings and dynamic powers of the teachers and traditions, not simply outer marketing and personality adulation. They are a call for us to go deeper into our own sadhana, following a blue print that covers the whole of life in harmony with both nature and the spirit. May the reader connect to the ancient wisdom traditions as part of the greater transformative movement of consciousness, without which our lives remain at the surface.

DR. DAVID FRAWLEY
(Pandit Vamadeva Shastri)
Author, *Yoga and Ayurveda,*
Inner Tantric Yoga, Yoga and the Sacred Fire

Introduction

Veda means knowledge in Sanskrit, and the ancient Vedic tradition associated with India is firstly about knowledge. It is not about any kind of knowledge, but it is particularly about knowledge as a means of understanding and experiencing reality. Therefore it is a tradition that does not tie itself down to any particular system. Instead it has an open and inclusive view, which also regards a rational approach as essential. However it does not limit itself to rationality alone, but it also considers an approach based around inner perception, as this forms an important part of our understanding and experience of reality.

'A Vedic Life' is a presentation of some of the important features of the Vedic approach to knowledge. It does this not for any particular religious or sectarian reason, but for the purpose of knowledge as described above, and so it is hoped that it will be able to offer something to a variety of viewpoints. As mentioned, the Vedic approach to knowledge is not restricted to any particular person or book, though it is closely associated with the Vedas (books of knowledge) as they represent its open and inclusive approach to knowledge. This nature of the Vedic texts is reflected in the following verse from the Rig Veda, a primary Vedic text, 'reality is one, but the wise call it by many names'.[1]

The Vedic tradition is not unique in its open approach to knowledge, but it is perhaps the deepest and oldest such system. It is also the best preserved, though much of its knowledge has been lost today. Obviously the system of modern science also aspires to the same goals, however it does not really recognise thinking other than that which is based on rationality and sensory data despite many of its pioneering thinkers having a significantly different view. This book will look at the strengths and weaknesses of the approach presently adopted by modern science, especially in comparison to the Vedic approach to science.

As Sanskrit is the main language of the Vedic tradition, much of the terminology used in the book is in Sanskrit. Again, this is not for any ritualistic reason, but because the Vedic language Sanskrit has a highly developed vocabulary in connection to psychology and philosophy. Additionally, an attempt will be made to demonstrate that Sanskrit is a natural language for the purposes of knowledge.

The book will mainly adopt the perspective of a rational presentation (as opposed to a more poetic or metaphoric approach) to demonstrate the rational approach of the Vedic tradition, and also because it is particularly suitable to modern thinking. The Vedic approach is thus not only ancient, but also modern and futuristic. This is especially so when we consider the systems of knowledge it uses, and its perspective on science as an art, and art as a science. We shall see that it is a natural system much like rationality itself, and in this regard it gives nature and a natural approach primacy. Thus it is able to draw on nature's vast and infinite qualities, in contrast to an approach that places humans first. The importance and necessity of this will be one of the main ideas put forward.

The first part of the book relates more to the universal factors of philosophy and cosmology. It forms the basis of the second part, which is an application of the first part at the level of the individual. The first part is like a commentary on the Vedic system of philosophy as presented in the Upanishads, the philosophical section of the Vedas. Though it is abstract at times, it is important as a foundation for the more practical second part, which includes areas such as science, art, medicine and self-investigation. The Vedic tradition is eventually about practical knowledge, or being one with knowledge, rather than just intellectualism. As such the two parts of the book must overlap due to the natural identity of the universal with the individual. It is a state of completeness, contentment and happiness, which is ultimately the goal of knowledge.

Finally, like a good astrologer, we can all see into the past and predict the future, especially the most important parts: we were not, and we will be not. Relatively speaking the rest should be trivial detail. However the consequences of the rest are obviously extremely important from our perspective, and we shall attempt to examine some of these. Such an examination is particularly relevant now as there appears to be a general increase in outer activity, with a corresponding decline in our concentration skills. Similarly, there also appears to be an increase in depression accompanied by an increase in our consumption levels. The book aims to return to our fundamental and natural strengths in order to try and overcome these trends.

ENDNOTES

1. Rig Veda 1.164.46

Part I: The Universal (Brahmanda)

This is Infinite/Complete, that is Infinite/Complete,
from the Infinite/Complete comes the Infinite/Complete.
Take the Infinite/Complete from the Infinite/Complete,
and still Infinity/Completeness remains.

Peace Mantra of the Isha and
Brihadaranyaka Upanishads

1.

Vedanta : Philosophy and Metaphysics

There was neither Existence nor Non-Existence.
Rig Veda 10.129.1

Vedic Science and Modern Science

The Vedas refer to the main literature and knowledge of ancient India, and its philosophical sections are called the Upanishads. The philosophy of the Upanishads is known as Vedanta, and as the implications of this philosophy are the underlying principle of all the factors that comprise a Vedic life, it is the subject of the first chapter. One of the purposes of this book is to bring out the importance of the natural and rational approach within the Vedas and Vedanta, plus the various disciplines that are based on them. If something is obviously irrational, it cannot be true, and thus it cannot be a part of the Vedic system whose primary goal is establishing the truth or reality. On the other hand, another purpose of this book is also to demonstrate the weakness of relying on rationality alone. Therefore this book also looks at the other means of knowledge that are available, and the approach of the Vedic system towards this.

As a rationale involves a scientific approach, it is appropriate to examine the meaning and practice of the words science and rationality first, when examining the philosophy of Vedanta. Science can be described as the establishing of truth or reality at a fundamental level, and the rejection of what is false. Rationality is one tool of the scientific way, but we appear to have no fundamental rationale for rationality. The only seeming rationale we have is that it appears to work according to our requirements. This is fundamentally arbitrary, and it is much like accepting rationality on faith as the best way to proceed. It is similar to a desirable ethical principle such as minimising harm. In fact, by definition there cannot be a rationale for rationality, for it would be self-referential as rationality does not contain its own proof, and therefore it would have no fundamental basis to stand on.[1] It indicates the incompleteness or finiteness of rationality alone as a system of knowledge. This will be explored further below, and also in the chapter on Vedic Sciences.

An important point to note is that the above definition of science is also the goal of Yoga and the Vedic sciences, and all of the Vedic tradition (which is now also referred to as Hinduism). The main difference between modern science and these older sciences is that they have different perspectives on what they consider fundamental, and hence they have a different priorities as far as methods of establishing reality are concerned. Modern science currently relies on data input from the senses, along with its statistical and logical manipulation, as the primary method of establishing reality. This seems naturally attractive as sensory data and the laws it obeys seem to be the most stable or fundamental things we have at our disposal, plus they are easily accessible. To this extent it has had spectacular results at this level of sensory data. However, an important weakness with this is highlighted when we consider our role as the scientist, or observer, as explained below.

Self-Reference and the Requirement of a Subject and an Object

To be a true observer or scientist at a fundamental level, we must accept a subject and object duality between us and the data we observe. If we do not do this then there is no difference between the subject and object, and the scientific system becomes self-referential. It cannot be objective as it has no separate object to be objective towards. It will always appear to come to boundaries that it cannot cross as far as its inquiry goes, or it will start obtaining nonsensical results. The boundary is created by a self-referential or circular loop from which there is no basis to step out. Modern science appears to be going down this path, as it defines the observer to be exactly the same in structure and system as the observed data. It uses its sensory data approach as primary when considering either of them. As a result it has been much less successful at a fundamental level in contrast to its spectacular results at an outer sensory data level, because it has been unable to answer the fundamental 'why' of things since the beginning of its methodology. This appears to be beyond its natural boundary. It has also come across paradoxes or nonsensical results due to self-reference, as noticed in the field of Quantum Mechanics, Logical Philosophy[2] etc.

Vedic science in contrast to modern science, does postulate a subject and object duality. According to it the attraction of sensory data for science must pre-suppose the existence of an observer in order not to be nonsensical or limited. It may appear to be a more complex approach, but it is a more natural one as we shall see. It requires the observer to have some non-sensory component at least, in order to be distinguished as the subject of the sense based objects. Additionally, the permanence or stability of such an observer is greater than sensory data, as sensory data is dependent on the senses and memory of the observer, whereas the sense of existence of the observer is more fundamental than that. Our feeling of our existence appears to precede the action of our senses, logic and thoughts generally, and thus also our memories. It is the most natural and fundamental awareness we can have, and such a method of perceiving the nature of reality is called direct perception. It will be examined through the course of the book as a means of obtaining knowledge, especially in the chapter on Vedic Sciences. It is considered superior to rational thought and sensory perception due to the fact it goes deeper than them in obtaining knowledge, and in fact it is the means by which the eternal truth of principles such as rationality, ethics etc. is realised. This greater permanence or stability of the observer is therefore a more natural and superior basis for an inquiry into reality than sensory data.

A further consideration in this regard is that objects only take on a meaningful existence as long as there is an independent subject or observer to detect them. Without a subject there is no way of defining an object, as there is no detector to conceive it. In fact, an object can only be defined relative to an independent and contrasting subject. In addition to these considerations when we add the fact that the subject (ourselves) is the reason for any scientific inquiry, subjectivity naturally should be given primary importance in any model of science.

Vedic science therefore chooses to base its primary method of establishing reality first and foremost on an inquiry into the existence of the observer, or self-inquiry, and so it concerns itself more with inner reality and less with sensory reality. The philosophical establishment of the primacy of the self is known as philosophy of Vedanta, a part of the ancient Vedic literature and thought of India. As we shall see next, for Vedanta to make sense at a fundamental level, there can be only one self, and this Self thus becomes the Self of all.

Advaita Vedanta

So the awareness of the self as 'I am', or in Sanskrit 'Aham', or the more abstract 'Om', is the basis for an inquiry into fundamental reality and all thought, from the perspective of Vedic science. We saw a reason for this was because it is considered to be the first, most enduring, and invariable awareness that we have. It can be argued that certain sensory objects outlive humans and so are more enduring. However no object is stable for any finite amount of time, even by the theories of modern science. Objects appear and re-appear at the speed of light by these theories. To this extent it is questionable as to what actually exists as far as sensory data is concerned. With human awareness these interruptions do not appear to exist, though there are states where consciousness is considered to be lost. Careful consideration reveals that actually there is actually never a loss of self-awareness, only changes in the field of consciousness, i.e. from a waking state to a dream state, or even to dreamless sleep and coma-like states. Even in these last two there is still self-awareness, as we are aware of having been in those states upon re-entering the waking state. If there was true loss of awareness, there would be no knowledge of having been in those states. A possible objection to this is that there may also be a loss of memory of having been in these states. However once again, there is even an awareness of having a loss of memory, so there must have been self-awareness at the time too.

Thus the existence of a non-sensory subject as the self at a fundamental level is the basic premise of the Vedanta philosophy. So we see that Vedanta begins with a rational approach, but takes it further via direct perception. Additionally, Vedanta has insights which show that in reality all the selves of all the subjects are the one and the same Self. We are able to naturally acknowledge that 'something' exists at a fundamental level due to the recognition of the existence of our self. The next step is to examine what it means to 'exist'. Here the insight is that in order for something to exist, i.e. to be true or real, it must always exist or always be true. There cannot be an independent or fundamental state that is true sometimes and not true at other times, or that is true up to a certain point in time and false thereafter. From this we have to conclude that fundamental existence must lie outside time (transcendental), and therefore it must be beyond changes and qualities. This in turn implies that *existence or reality cannot be more than one thing* at a fundamental level. For if there is more than one thing, these 'things' must be different to each other in some way. Differences or changes require time in order to occur, which means these 'things' must also exist in time. It follows from this that they cannot be eternal or fundamental.

This is the summary of the philosophy of advaita Vedanta, or the philosophy of undivided (non-dual) fundamental existence. It is sometimes also referred to as the philosophy of monistic idealism or monistic realism. It gives us the important result that directly perceived truths such as the Self or intelligent principles such as rationality, must all be eventually one and the same, and must also be eternal. An important point to note regarding advaita is that it is complete in itself, and it is free from any arbitrary postulations which are always subject to revisions and corrections. It is based on direct perception and rationality only, and as such, it is a philosophy that can be described as being naturally true.

Maya or Pseudo-reality

We can notice that the monism of advaita also causes the subject and object duality to collapse, but with important differences from the view point of modern science. In advaita it occurs at a transcendental level of completeness, as opposed to a material or sensory level. Due to the self complete nature of transcendence, it does not require the duality of a subject and object as we shall see. However more importantly, advaita acknowledges this by the startling claim that all of the individual selves, and in fact all the mental and physical variety we sense as manifestation cannot exist at a fundamental level. Their pseudo-existence, which is the concept of Maya, somehow must be contained at the fundamental level as we shall see in the next section on the dvaita philosophy.

Examples of the Harmony of Advaita

As the advaita philosophy is naturally true, we can also expect it to naturally resolve areas which are otherwise unexplainable. For example it gives us a fundamental rationale for non-violence, kindness and other principles of ethics, which otherwise appear to be only arbitrarily attractive, i.e. which otherwise are more or less accepted on faith only. In fact in materialistic philosophies there can be no fundamental rationale for ethics, as matter itself has no principles. In monism harming others is the same as attempting to harm yourself at a fundamental level. (Though reality itself cannot be harmed due to its transcendental nature.) There can be no better rationale than this for the principles of ethics, and we can see that we find these principles naturally or intuitively attractive, as they are based upon the unifying reality of our own nature.

Another issue addressed is that if we have the commonly accepted scenario where there are fundamentally many selves as subjects, we have the difficulty that we are unable to identify and establish a general model of reality. Each self will detect reality differently to a certain extent, by virtue of that self being different from the others. (Otherwise it would not be differentiated as a separate self.) It raises the problem of there being no two same 'realities', and this makes it impossible to know if there is even a single reality, as all these 'realities' are relative to one another with no fundamental basis.

The monism of advaita also puts into perspective the many Gods, or more accurately the related concept of Devas, that exist within the Vedic tradition. If Gods exist with qualities or differences in time, we have seen that they too ultimately cannot exist at a fundamental level. This makes advaita an atheistic philosophy in the conventional sense, yet it allows for awareness and consciousness. These points will be looked at further in the next chapter on Cosmology, as well as the chapter on Devas.

Advaita and Dvaita

However, the advaita philosophy does also leave us with a seemingly unanswerable question from the perspective of rationality. It establishes all existence or reality as a transcendental unity with awareness and consciousness (known as Brahman in Sanskrit), but in contrast, our experience obviously also establishes all the diversity of mental and physical sensory manifestation, or Maya. There are approaches which reconcile these positions, and these positions must naturally be reconcilable as they are naturally derived.

A common but incomplete method to deal with the above question is approach of the philosophy of dvaita (division or duality), with which Vedanta is also associated. Here a second principle is postulated (Prakriti) as the object for the subjectivity of Brahman to allow for manifestation and its variety. Prakriti's existence has to be dependent on Brahman (more commonly referred to as Purusha in this case),[3] as Brahman is the only fundamental reality. Through this dependency Prakriti gets the ability to exist eternally, but it also has the ability to keep changing, which allows for the variety of manifestation. Unlike advaita, it is apparent that dvaita is not a natural system of philosophy, as it involves the arbitrary postulation of the second principle of Prakriti. Due to this fact, it is possible to establish a weakness in it which would take it back to the natural position of advaita. The weakness is that in order for Prakriti to actually exist, it must have an unchangeable factor, i.e. a factor that lies outside time for the reasons already outlined. This would in fact make it the same as Brahman, and the philosophy of advaita would be re-established.

The approach of advaita on the unanswerable question is that it must inevitably or naturally arise due to the inevitability of advaita. However it also naturally does not exist as a question, and as a result it is not only unanswerable, but it is not required to be answered. The question, and the circumstances in order to have the question, fundamentally do not exist as they are also pseudo-real or Maya. Therefore the question cannot have a real answer. Another way to look at this is that the questioner and the question are both subject to the passage of time. The answer required must also be subject to time. The question itself concerns Brahman which lies beyond time. For reasons of self-reference or circularity again, it becomes impossible to answer such a question. The purpose of this type of advaitic reasoning is to highlight the limitations of rationality, and to emphasise that self evident truths such as our self existence require a self-evident or self-proving systems of knowledge like that of direct perception. In the second part of the book we shall see how the practices of Yoga and Tantra overcome these limitations at a more practical level.

The natural conclusion of this is that all of manifestation is in fact not what it seems, but there is only Brahman. For a transcendental 'reason' manifestation appears to us as something other than Brahman. This 'reason', which must be Brahman, cannot actually be known, described, or even actually counted as one thing, because there is no second thing by which Brahman can be known by, defined against, or counted from. However, it is naturally possible to be only Brahman and its knowledge of all, via the method of direct perception, as we shall see later. Direct perception ultimately transcends all dualities and manifestation, but this level of direct perception is extremely difficult to attain, and thus from the perspective of advaita it is essential to begin with the dvaitic approach first.

Dvaita has a weakness as we have seen, but it is a necessary weakness due to the pseudo-real nature of manifestation. It gives legitimacy to the rational or scientific (modern and

Vedic) approach by establishing the necessary dual of the subject and object. It is for this reason that principles such as rationality and ethics are considered to emerge from Prakriti, as they become possible only after the duality of Prakriti is established. Thereafter by progressive thought and insight, Prakriti realises the limitations of itself as it begins to discriminate between the undivided reality of Brahman and the pseudo-real variety of manifestation. In this manner Prakriti sets the framework for the negation of itself, or for realising transcendental reality as itself, which leaves only direct perception and Brahman. The various steps of this progression underlie the thinking behind Vedic cosmology, Yoga and the other Vedic sciences, and these will be described later in the book.

The three main Yogic paths are usually described as Karma (action) Yoga, Bhakta (devotion/desire) Yoga, and Jnana (knowledge) Yoga. The above approach of advaita/dvaita combines all three, as it requires action or scientific effort which leads to gaining knowledge, and this is propelled by a desire to rest in the natural and complete state of the Self or Brahman. Existence as other than Brahman is illusory and unnatural at a fundamental level, and therefore it is incomplete and unsatisfying.

The Shad Darshana or Six Philosophies

The Vedic system of philosophy has several stages before it culminates with philosophy of advaita Vedanta. There are six major such stages which collectively are know as Shad Darshana, or six 'points of view'. The first two stages are called Nyaya and Vaisheshika which relate to philosophy of logical thought and atomic theory of matter respectively. As they are dominated by logic and matter, they are mainly materialistic philosophies with an underlying transcendental element. Their outward focus makes them a natural starting point in the Vedic system of philosophy and science. They are much like modern science in approach, though they are less developed materialistically as they were recognised as being just the first stages of any philosophy. However, their materialistic aspect made them popular with the atheistic and materialistic thinkers connected to the Vedic system of thought in the past.

The next stage in the progression of Vedic philosophy is the system of Sankhya.

Here thought processes are naturally drawn inwards from the previous outer stages, by acknowledging direct perception and the existence of the 'I am' awareness. To this extent it establishes the dvaita or dualistic philosophy described above. It also enumerates (Sankhya mean enumeration) principles for a model of cosmic evolution as we shall see in the next chapter, and this also forms the basis for the Yoga and Vedanta systems. After Sankhya comes the stage of Classical Yoga, which is mainly a practical system concerned with developing direct perception via its techniques of meditation. It should be noted that the Vedic system emphasises practical knowledge, because until knowledge is integrated as a part of our experience it remains apart from our self. A monistic system requires integrated knowledge or wisdom as the most natural and fulfilled state, and not intellectual and factual knowledge apart from ourselves. Hence the practice of Yoga is an important aspect of the otherwise abstract advaita philosophy too.

The final two stages of Vedic philosophy are associated with advaita Vedanta. The first is a set of esoteric inner rituals which aim to take the mind to states beyond rationality in order to facilitate direct perception. The second is the philosophy of advaita as already described. It is the culmination of the six points of view as it recognises the existence of reality only.

The Vedas and Upanishads

There are four main Vedas, the oldest and most important of which is the Rig Veda. It is the oldest known text to humanity, and from the astronomical and other information contained in it, it appears to have been compiled over a period of a few millennia beginning about 10,000 years ago.[4] The version we have today was probably completed about seven to eight thousand year ago. It was originally maintained and transmitted in the Sanskrit language orally only, i.e. it was maintained via memory. There is no clear picture when the written version of it first appeared, but it is very likely to have been several millennia after the oral version. The method of oral transmission is probably why it has been preserved so accurately, as it has been constantly recited and scrutinised. This is difficult to do when texts are put onto paper as it allows for errors to go undetected, and also to increase over time. In addition, the oral tradition came with many sophisticated mathematical error detection techniques,[5] where errors are easily detected on recital. The accurate preservation was given so much importance to the extent that not only the accuracy of every word, but the accuracy of the pronunciation of every syllable was considered extremely important. Considering the main text is of about 10,000 verses this was a considerable feat. The reason that the correct pronunciation was given so much emphasis was because it was believed that the sounds had a positive vibratory influence on the mind in addition to their intellectual meaning, or that they had mantra power. The power of mantras and the Sanskrit language will be addressed in the chapter on Mantras.

The Rig Veda is considered to have many levels of meaning,[6] and is a highly poetical and metaphorical work concerning existence and nature. It is said to be the result of the inspiration, or direct perception, of many sages or Seers (Rishis) over several millennia. As we shall see, it is a seminal work across the Vedic system of thought, and indeed human thought, as befits such a carefully constructed and preserved text. The final sections of the four Vedas (in addition to their main texts) are known as the principle Upanishads, which date from about five thousand to three thousand years ago.[7] The Upanishads tend to be a little less metaphorical, but more philosophical and intellectual than the main Vedic texts. The Vedanta philosophy is usually associated with them as it can be first directly seen there, though it is also present in the main texts in a more subtle form.[8] The poetical nature of the main Rig Vedic text encourages direct perception, and it was felt within the Vedic tradition that this ability was beginning to decline several millennia ago, as will be explained in the yuga (great cycles of time) section of the chapter on Vedic Astrology. Hence the Upanishads were added as the more intellectual synopsis of the Vedas. In later times, when the ability of direct perception decreased further still, the Puranas became more prominent. In addition to the intellect, they also appeal to the emotional nature. The Upanishads and Puranas begin to look at cosmology more explicitly, and this takes us to the subject of the next chapter.

ENDNOTES

1. To see this demonstrated in Logical Philosophy, refer to the work of Kurt Gödel as set out in The Emperor's New Mind by Roger Penrose. Amit Goswami's Self Aware Universe is a good introduction to Quantum Mechanics where this point is also relevant.
2. As above.
3. As these terms refer to transcendent 'concepts', there is generally some variation in their usage. Here we have taken Brahman and Purusha to be equivalent, but sometimes Brahman is referred to as the supreme reality beyond existence and non-existence, whereas Purusha relates to the reality responsible for existence. Purusha is thus connected to manifestation and is closer in meaning to Ishvara, which will be introduced in the next chapter.
4. For further information see God, Sages and Kings and other works by David Frawley.
5. Visit the SAKSI website www.vedah.com for more on this.
6. This is examined in Secret of the Veda by Sri Aurobindo and Wisdom of the Ancient Seers by David Frawley.
7. As with the dating of the main Vedic texts, there is some debate as to the antiquity of the Upanishads, with dates usually ranging from two to five thousand years ago.
8. Please refer to The Vedic Vision of The Upanishads by David Frawley, for a detailed explanation of this.

2.

Sankhya/Itihasa: Cosmology, Evolution and History

All this is One Reality.

Chandogya Upanishad 3.14.1

Sankhya and the Concept of the Triad

From the chapter on Vedanta above, we saw that manifestation is only pseudo real, and it requires a subject and object, or the dvaita (dualistic) aspect of the Vedanta philosophy. The Sankhya[1] philosophy is dualistic, and is the prime philosophy in Vedic thought to explain the cosmology of manifestation and evolution. Sankhya is known for its numerical list of 24 tattvas or principles, plus Purusha, as the foundation of its system. In this chapter an attempt will be made to show how these principles can be naturally derived, or are self-selecting, within the framework of dualism.

We have already seen the derivation of the two most important principles, Purusha and Prakriti as the universal subject and object, in the previous chapter. In order to derive the remaining principles, along with their cosmology, we shall begin with the establishment of the fundamental concept of the triad.[2]

It is possible to see naturally via direct perception (much like it is possible to see the truth of logic or ethics) that all systems aim for a state of equilibrium, and further, that this state requires at least three points, or a three-fold nature.

For instance, a needle will not balance on its point, but will fall flat in order to achieve a position of equilibrium. A one or two pointed object is not stable, but when we add the third point, it becomes naturally stable as a tripod. Geometrically, a triangle is the shape with the least number of corner points (three) that achieves closure. A circle achieves closure with no corner points, but it has curvature instead, which can be equated to an infinite number of corner points. Hence a circle is more appropriately used as a symbol for transcendental closure. We can also think of the fundamental completeness of the triad as the three complementary states of active, inactive, and the balance between the two, which is the basis of the gunas (universal qualities) described below. Similarly, we have catabolism, metabolism and anabolism, which relate to the doshas of Ayurveda, as we shall see.

Another option for stability is obviously the unity of advaita, but this automatically becomes transcendental, or beyond manifestation, as it is all encompassing. There is nothing else in a 'unity', including the concept of numbers and counting. It is the real stability according to advaita, with the triad being a pseudo-stable state. When there is a duality (or more) of manifestation, a polarity as the difference between the dual is automatically established. This causes an urge towards the stability of the unity, by the principle that any unstable system tries to find a natural equilibrium.

So we have the choices for stability as a transcendental unity, or a triad in manifestation, i.e. according to the level of our direct perception, existence is either a real and transcendental unity, or an apparent triad in manifestation. As the triad in manifestation represents our perspective, we shall examine this further next.

The Roles of Prana and the Three Gunas

This triad must initially arise out of the dual of Purusha and Prakriti. A subtle point to observe here is that a triplicity is established from a dual state by the polarity (i.e. an attraction or tension) between the pair forming the dual. When the attraction ceases or is satisfied, a unity is established, and while it exists, it is the force of life that is responsible for manifestation. Between Purusha and Prakriti, this force is known as cosmic Prana, which is sometimes explained as the will[3] of Purusha or Brahman. It keeps the cosmic dual separated, which enables the formation of a stable triad. As such, Prana is the force behind manifestation and life. Also, as the force between the cosmic couple of Purusha and Prakriti, it forms the basis of Tantra[4] as we shall see later in the book. We should note that the cosmic triad lies outside time, so any discussion such as the above must be via metaphors, and this also applies to Brahman in the first chapter.

In the Sankhya philosophy of cosmology, the cosmic triad is reflected as the three gunas. These are known as Sattva guna (balance), Rajas guna (activity), and Tamas guna (inactivity). Initially the balance of unity of Purusha is known as pure Sattva. Then Purusha, Prakriti and Prana are balanced by Prana as the activity of the force between Purusha (which is balance, but inactive as the sole reality) and Prakriti (which is the field of activity as the basis of manifestation, but of an inactive nature). In analogous fashion, the balancing action of Prana acts as a lower Sattva for the subsequent balance of further triads of manifestation. Thus the three gunas form a complement, and are foundational for manifestation. Sattva is the reality of manifestation, Tamas is the basis of manifestation, and Rajas is how reality and manifestation connect to each other.

So we can see that the gunas form the framework for all the other triads of mental and physical manifestation to be conceived by the transcendental state of consciousness, or Purusha. Prime amongst the other triads that are formed is the one associated with the materials (mental and physical) of manifestation. The dual here is often symbolised as fire and water, with the polarity or motion symbolised by air. In the chapter on Ayurveda we shall see how this has practical applications from the perspective of day to day living.

It is therefore evident why the concept of the triad naturally occurs across cultures and through history, right up to the science of the present day. Apart from the Vedic tradition, it is the present in the Chinese system as Yin, Yang and Chi, and in the European system as the Father, Son and Holy Ghost. In modern science it is seen in the form of the electron, proton and neutron. Obviously these are only just a few examples.

The Rig Veda and the Triad

The mechanics of the triad in cosmology can be more easily understood when we consider the triad as symbolised by fire, air and water, particularly from the perspective of the Rig Vedic triad of Agni (cosmic fire), Vayu (cosmic air) and Soma (cosmic water). Cosmic fire is used to symbolise all the energy in the cosmos at the physical and mental level, including the intelligent laws, force, light, time latent in it. An analogous symbology is true for cosmic water and cosmic air.

The connection and complementarity of this triad is evident from the image of the fire of Agni burning the fuel of Soma, and producing the resultant vapour, or motion, of Vayu. If the fuel is 'clean', a pure fire is produced, with a clear and light vapour which rises upwards. The triad becomes an upward pointing triangle of the evolutionary process, and this is the goal of the practice of Yoga. Similarly, a 'dirty' fuel produces a toxic and dense smoke which descends, and the triad becomes a downward pointing triangle of involution or manifestation. In this imagery, Vayu represents the motion born of the couple, and is like Prana. However, we have just seen in the section above that Vayu or Prana as a unity can also act as the creative principle for a new triad (either upwards or downwards), and as such, can give birth to Agni and Soma in an apparent reverse process.

It is much like how the unity of Purusha gives birth to all manifestation.

Thus, whether it is the downward motion of manifestation, or the upward motion of evolution, they are caused by a disturbance of an equilibrium by energy to another level of equilibrium. The disturbance is caused by Prana, or Agni in the Rig Vedic symbology, and the ascent or descent itself can generally occur in three main ways as a reflection of the triad in nature. For example, it can be step by step like going up or down a mountain path, or it can be like the gush of a stream or the gust of the wind, or it can be instantaneous like a flash of lightening. This is reflected in our psychology, and will be examined in the second part of the book.

The Relativity of Morality

The terms 'clean' and 'dirty' fuels are relative terms, and represent either the projection of consciousness downwards towards denser manifestation, like a dream, or the return upwards to the transcendental as true existence. Fuels that promote real or true existence can be termed as relatively clean, and fuels that project manifestation can be termed as relatively dirty. The term dirty, however, does not apply to the process of manifestation of nature by the will of the consciousness of Purusha described above. This form of manifestation has transcendence underlying it, and is beyond the idea of clean and dirty.

As a result, a fuel that is dirty at a particular stage can be clean when compared to a lower stage. This is an important point as far as social conduct is concerned, for it means the merits of behaviour codes and practices are relative. This also means the concept of goodness in religions and religious practices is also relative, i.e. it depends on the individual circumstances. The concept of clean and dirty fuels will be further developed in the subsequent chapters of the book.

Agni, Vayu and Soma thus are the building blocks of mental and physical manifestation, and so they are also the basis of the structure of the human body in the forms of the doshas, which we shall see in the chapter on Ayurveda.

Cosmology of Intelligence, Force, Light and Time

We have seen that cosmic Prana, which arises from Purusha and its creative aspect or cosmic dual Prakriti, is the force behind all manifestation, and it arises outside time. A force acts as a potential for energy, and thus from the cosmic Prana arises a cosmic light or energy, also lying outside time, known as Prakasha. From this light we get the basis for the 'manifestation' of physical time. It is also from this energy that the triads of primal matter and physical matter are established. These points can also be seen from the results of modern science, i.e. that light defines time, and the equivalence of matter and energy.

Strictly speaking, we cannot extend the results of the logic of modern science to the transcendent field in this manner due to the infinite nature of transcendence. However, from our perspective which lies within time and is based on rationality, we can extend these results as an analogy in order to give us general guidelines. Additionally, the results of modern science cannot only be dependent on the finiteness of rationality, but must ultimately involve a certain amount of deeper perception, such as the principle of the truth of rationality. This is a perception into the eternal principles that are necessary for any system of thought, and it allows us to get a sense of the transcendental even in the rational.

Manifestation in time is therefore dependent on Prakasha, and we can see that all manifestation is also pervaded with eternal and intelligent principles and laws, such as rationality and the law of cause and effect. Prakriti itself requires these laws and principles for its purpose of manifestation, and foremost amongst these is the law of karma (relating to the law of cause and effect) which will be examined in the second part of the book. As such, these principles must be seeded in Prakriti, and this cosmic intelligence inherent in Prakriti is known as Mahat in Sanskrit. *These factors establish the fact that even the materials of physical manifestation are permeated not only by force, energy and matter as per modern science, but even by time and intelligence.* If there is a particle of matter, then we also have the existence of energy, force, time and intelligent laws such as logic or the law of cause and effect, as its properties.[5] This is why we notice that the intelligent and rational results of mathematics, howsoever abstract, always have an application at the physical level too.

Modern science does recognise this to a certain extent on the physical level by connecting light with time, and light as energy with matter, but it does not push through the results of these ideas. Some of these ideas are being forced upon modern science in the field of Quantum Mechanics.[6] The Vedic sciences do take these ideas further, as we shall in the chapters on Vedic Astrology and Yoga.

The Pentuple

The transcendental triad of Purusha, Prakriti and Prana is complete, as Purusha is the underlying unified state of all, with awareness and intelligence. With Prana and the resultant Prakasha, the process of manifestation begins as a field of consciousness and the three gunas. Prakasha forms the basis of a lower unity, which once again forms a triad based on a lower Prana as a polarity. The action of Prakasha is responsible for time, and hence this lower triad functions in time (albeit still not physical), and is no longer transcendental. This process is repeated to the physical level as a series of downward triangles, and is the process of involution.

In this involuted state, qualities and changes become manifest due to the presence of time. (Without any changes or differences we cannot mark the passage of time.) This differentiation causes a process of individualisation to occur. The lower triads are no longer complete as they are just an image of Purusha and its unity and intelligence. These triads lose awareness of their underlying transcendental basis, and hence they are said to enter a state of ignorance or avidya. This is equivalent to an individualised version of the universal illusion of Maya.

Therefore, to complete the triads that exist in time, a pentuple state with two additional factors is required.[7] The first factor is a reflection of the Self as the individualised ego or self known as ahamkara (I maker or individualisation). The second factor is the reflection of the cosmic intelligence, Mahat, as an individualised intelligence known as buddhi. The pentuple becomes individually active after the manifestation of the five elements during the cosmological process, as will be explained in the next sections. It should be noted that the fourth and fifth factors of the pentuple are not required for stability, but are necessary for completeness which is required for manifestation in time. The triad component of the pentuple is still the key building block for the process of involution, due to its stability.

The 24 Principles or Tattvas of Sankhya

The pentuple therefore forms the basis of the Vedic cosmology in time. The intelligent consciousness of Purusha projects its manifestation in time as a pentuple via Prana and Prakasha. The first stage of this pentuple projection is the five tanmatras or mental senses in the field of consciousness. These five are the analogue of the physical senses, and create all the mental worlds from emotions to creativity and inspiration through the kingdoms of nature, as we shall see. From the tanmatras the five physical sense organs are formed to detect five different sensations. The five different sensations are provided by the five great elements which complement the senses, and they make up the entire physical universe.[8] We have seen that as far as the building blocks of manifestation are concerned, these pentuples must relate to triads. The mechanism behind this will be looked at in the chapter on Ayurveda.

The densest two of the five elements are earth (solid) and water (liquid). Together they make up the non-gaseous matter aspect of modern science. The next element is fire (energy), which is obviously the energy aspect of modern science. The next element is air (gas) which completes the mass aspect of modern science, along with the first two elements. Finally there is the space element which is the container of all the other elements, and thus it is considered the first of the elements. The space element is not fully appreciated by modern science, but it has a mathematical equivalent as a force field. Also in mathematics, we can say it is also equivalent to the number 0, perhaps the most important number.

The five elements correspond to the senses in the following manner:

Hearing – Space
Feeling – Air
Seeing – Fire
Tasting – Water
Smelling – Earth

The tanmatras are the mental equivalent of the five elements. We can understand the order and correspondences in the above table when we consider them from their perspective. Hearing in this case means the 'hearing' of silent 'words' as mental vibrations, and these words are the basis of the imagery of the other mental elements. Hence they form a container for the other mental elements, much like physical space contains the other physical elements. Thus the first element of space is associated with hearing.

The first thing that is possible in the empty space of the mind is the motion of consciousness, and this is symbolised by air physically. In the physical world the motion of air is the least dense form of motion, and so closest to space. Mentally, it is only when the silent words move as thoughts that they provide the imagery for the rest of the elements. We detect motion as touch or feeling. When there is no motion, an existing contact or touch cannot be recognised. For example, when we lie on a surface without any motion, our body will become unaware of the contact with the surface (provided there is no exchange of heat, or feeling of hot or cold, which is the next element).

Energy requires motion to manifest, and as soon as there is motion, there is energy. Hence these two elements can be considered to be formed simultaneously. Mentally thoughts form moving words, but are given an actual image only by the energy of light. We know from modern science that all forms of energy are equivalent to light energy. The method of perception for this is obviously via sight.

The level of energy manifested in motion depends on the level of densification of the motion. The less energy the more dense, so the next element after the most subtle motion of air is the more dense water, and finally the most dense earth. Water brings in its own 'touch' as taste, but is yet odourless. The mineral, or earth, element is responsible for smell in solids, liquids and gases. Thus we see the movement through the elements from space to earth as the progression from force (relating to space) to energy to matter, and this is also recognised by modern science.

Overall, space as the first physical projection of consciousness from light or Prakasha, contains all the other elements in latent form. On the other hand, the earth element as the final product of manifestation contains all the other elements in dense form. For example, solid matter has smell and taste, it has energy as heat, it has movement at a molecular level, and it occupies and contains space. In fact, most solids are overwhelmingly hollow according to modern science.

The five senses also correspond to five motor organs, in addition to the sense organs, in the following manner:

Hearing – Speech
Feeling – Hands
Seeing – Legs
Tasting – Urinary Organs
Smelling – Excretory Organs

Speech corresponds to hearing as it is the spoken word. Hands are the main instrument of feeling, which is the sense corresponding to the air element. The legs give mobility to the entire body, and the key sense needed for movement of the body is sight. Urination and excretion naturally relate to the water and earth elements and their corresponding sense organs.

Thus we have four groups of pentuples as:
Five Tanmatras
Five Elements
Five Sense Organs
Five Motor Organs

This gives us 20 of the principles or tattvas of Sankhya. In addition there are the tattvas of Prakriti, Mahat and the ahamkara as already described. The final tattva in order to complete the group of 24 tattvas of Sankhya is manas. This tattva is the level of mind created by the action of the tanmatras, and so it primarily corresponds to the sensate mind, and emotions such as anger, greed and fear. These emotions will be considered further in the chapter on Ayurveda. In its higher function, manas can also connect to the lower aspects of the individualised cosmic intelligence, or buddhi, via the rational mind. Purusha is not grouped in the main 24 Tattvas as it is not a factor of cosmology, but underlies the cosmological process as the 1st or 25th factor.

Eternal intelligent and rational principles such as the law of cause and effect are contained in this cosmology via Mahat and its individualised reflection of buddhi. Prana is contained in this system as the force between couples which form the triad. Light is contained too, as it arises from Prana, and it is the basis of the projection of the pentuple.

The Five Koshas (Coverings) and the Seven Levels

When the pentuple reaches the stage of physical manifestation of the five elements during the process of involution, it takes the form of five covers, or koshas.[9] They form the subtle constitution of all physical manifestation in the Vedic tradition, and they correspond to five levels of manifestation within the Sankhya system as follows:

1. Food Kosha - Five Elements
2. Prana Kosha - Prana as Instinctual Life-Force
3. Senses and Emotions (Manas) Kosha - Tanmatras
4. Intelligence as Direct Perception and Rationality (Vijnana) Kosha - Buddhi, which connects to Mahat.
5. Bliss, Causality and Direct Perception (Ananda) Kosha - Individualised Being or Jivatman as highest Ahamkara. Connects to the causal creativity of the cosmic triad.

The first three of these koshas form the triad of manifestation, and the fourth and fifth koshas are the reflection of Purusha as the Self, and its transcendental intelligence. The five koshas themselves form a triad of the three-fold soul, mind and body. The body naturally relates to the physical body, the mind to the manas kosha, and the soul to the bliss kosha. The instinct kosha connect the body to the mind, and the vijnana kosha connects the mind to the soul. In Vedic symbolism this three-fold form is expressed in many ways, such as earth (body), atmosphere (mind) and heaven (soul), or the waking state, dream state and deep sleep state. The waking state is where there is awareness of body, mind and potentially also the soul. In dreams there is awareness of mind and potentially the soul, but not the body. In deep sleep, if there is awareness, it can only be at the level of the soul.

To reflect completeness in time and beyond, the pentuple is expanded into seven levels[10] to include Purusha and its intelligence. The sixth level in this scheme is referred to Chit (pure intelligence), and the seventh level is referred to as Sat (pure being as the Self). The bliss kosha is the link between transcendence and time, and it is where completeness is thus first experienced. Hence it is referred to as the kosha of contentment or bliss. It is where Prakriti occurs, where the first triads are formed, where the process of manifestation and causation begins, and where time itself occurs. When it individualises as the jivatman, it is the basis for the conscious causation at the level of the individual, much like Prakriti is for the entire universe across all levels of manifestation. Chit, as the pure intelligence beyond creation, is reflected as the creative (but still transcendental) intelligence Mahat of Prakriti, as Prakriti is the creative dual of Purusha. Mahat then becomes intelligence in time as buddhi.

As Purusha is a unity, it is more accurate to refer to Chit as Purusha, rather than of Purusha. The same is true for Sat, and for an aspect of the fifth kosha or Ananda. Thus Purusha is Sat, Chit and Ananda, with Sat = Chit = Ananda. This gives the important result that being, intelligence and completeness or contentment are one and the same thing, and there cannot be one without the others. From this we can see how Purusha is the one seed for the concept of the triad, and also for the further reflected pentuples.

Ishvara, the Universal Being

Similarly, though each of the koshas of the pentuple appear divided or individualised by the ahamkaras, they in fact form one universal kosha. For example, though we feel we have separate bodies from one another, the matter that makes up our bodies knows no such boundaries. It freely moves from one being to another, whether it is other humans, animals, the earth, the air etc., behaving as one entity. We can also see that the ahamkaras at the level of the bliss kosha must unite as one as here they connect to the transcendental level. Our intelligence, emotions and instincts also unite into single greater koshas too. It is much like how we know that all the matter in our bodies is eventually connected within us as our one being, though its complexity makes us feel that it is an infinite number of separate parts. We also know that the physical matter in the universe has exactly the same relationship to other matter in the universe, as the matter in our bodies has to the other matter in our bodies. So the universal being, known as Ishvara in Sanskrit, forms the universal equivalent of our individual beings. From this we can see how the consequences of any actions belong to us all, whether the actions are conducted individually or collectively.

The Kingdoms of Nature and the Process of Evolution

The tattvas of Sankhya above describe the process of involution, or the descent of awareness, consciousness, force and light into mental and physical manifestation. Similarly, we can trace a process of evolution from physical matter to the state of complete stability of the unified consciousness of Purusha. This is ultimately the real equilibrium to which the pseudo-equilibrium of all manifestation must eventually return. The natural urge to return to the equilibrium of Purusha is known as dharma in Sanskrit, and the movement towards any equilibrium is due to the law of cause and effect, or karma. The concepts of karma and dharma, and the process of evolution generally, are the main focus of the second part of the book.

An implication of the Sankhya system is that individualised and evolutionary activity, or the evolutionary activity of the pentuples of the ahamkara, can only begin from the most involuted or physical state. If it were to begin earlier, the densest levels of manifestation would not form, as the ahamkaras would be naturally drawn to the real equilibrium of Purusha consciousness. The lower equilibriums of the triads can only occur via the will or force of Purusha to project manifestation, i.e. only by the unity of Purusha, and without any individualised activity. To this extent, when we generally use the term manifestation on its own, we shall consider evolution to be a part of it as a subsequent stage. However, the process of manifestation and evolution is obviously (even by our own experience) not uniform, but has its many ups and downs, as described by the analogy of clean and dirty fuels earlier.

Thus the individualised ahamkara becomes first activated (in the form of the five koshas and the septuple above) at the densest level of the five elements, i.e. where the earth element dominates the other elements. This is the mineral kingdom, and here the food kosha is most active of the koshas. The mind is dormant, Prana as the life-force of the instinct is barely evident, and the ahamkara appears more or less completely inert.

As evolution proceeds to the next densest level, we have the ahamkara manifesting as the plant kingdom of nature. Along with the earth element, the water element (corresponding to the plant sap) also becomes prominent. Also, the fire element begins to develop with the process of photosynthesis that commences in the plant kingdom. The Prana kosha begins to manifest as the plants begin to show basic instinctual properties. The next kingdom of nature to manifest is that of the animal kingdom. Here the air element becomes more prominent with the development of the lungs and respiratory system. The initial stages of the emotional mind and tanmatras begin to function with the development of the sense organs, and so the ahamkara begins to get activated in the field of emotions.

With the commencement of the human kingdom of nature, the space element becomes activated by the use of words and thoughts, and the non-physical aspect of the ahamkara becomes stronger. The emotions and tanmatras are further developed, and the initial stages of the buddhi or intelligence kosha are seen. A particular feature of the human kingdom is that it is where the process of involution (manifestation) and evolution particularly come into confrontation with each other. The ability to make decisions or buddhi is first developed here, and so both, correct and incorrect decisions (corresponding to clean and dirty fuels), can be made in this kingdom. As such, it is a kingdom of conflict. The positive is that the decision making ability of the human kingdom allows for a more rapid evolution compared to the lower kingdoms.

The last kingdom of nature that is possible is the Deva kingdom, as this brings in the fifth and final kosha. The Deva kingdom translates as the kingdom of light, and it is known as such due to its connection with cosmic light or Prakasha. It is the kingdom that links the world in time with the transcendental world, which is also the function of Prakasha. As a result it is mainly a non-physical kingdom, though through its transcendental connection it has the ability to manifest in any of the kingdoms. It can be divided into a lower and a higher part as below.

The lower part of the kingdom is the level above human kingdom that is subject to time, including the part of the intelligence kosha that lies above humanity. It consists of ahamkaras that are still evolving, albeit at a higher level than humans. As they advance, the ahamkaras begins to activate at the primary or causal level of manifestation associated with Prakriti, and the Self of the individualised selves (jivatman) begins to be realised. From this level the ahamkaras begin to merge into one as Purusha. (This is technically possible even from the lower kingdoms, particularly the human kingdom.) Advanced Yogis and Rishis (inspired Seers of knowledge or Veda) across all cultures and traditions, who also manifest in the human kingdom from time to time as its guides, are considered to comprise this level in the Vedic system.

The higher part of the kingdom is what is commonly referred to by the term Deva. It exists with the onset of time via the energy of Prakasha, and hence has eternal properties. Its is the basis by which manifestation in time occurs according to the Sankhya system, and it is also the basis by which the ahamkaras are able to evolve back to Purusha. It is the energy or light by which Prakriti is able to guide or control its manifestation in time. Thus, unlike the other kingdoms of nature which evolve to Purusha, it is not subject to evolution, and does not comprise of ahmakaras. It is perfectly evolved and real, being eternal, yet it is involved in manifestation, as a result of the same philosophical paradox seen with Prakriti in the chapter on Vedanta. It appears to be varied as many Devas by the many actions it initiates, and it is usually referred to in the plural, but it is really unified as the transcendent universal being, or Ishvara. In the western or Abrahamic tradition, it is referred to the kingdom of the Gods due to the God-like properties of the Devas, though this is not strictly accurate as the terms are not equivalent. Unlike the concept of God, the concept of a Deva is a natural function of manifestation, and it lies beyond general human intellect and qualities. As such, it can only be considered metaphorically and abstractly, or only directly experienced. The chapter on Devas will go into this in more detail.

As an example of the Devas, we can consider the role of Agni. We saw that the energy or fire element is responsible for the level of density of the other dense elements, i.e. whether they manifest as earth, water or air. Similarly, at a mental level, the level of fire in the tanmatras also determines the level of density or evolution of the tanmatras. In this manner, because the highest manifestation in time of this occurs in the Deva kingdom, the Devas associated with the fire element are the highest available support in order to facilitate evolution. Collectively, the evolutionary properties of the fire element at all levels unite as the cosmic fire, or

Agni as a Deva. We also saw this role of Agni in the cosmological process of the motion between higher and lower triangles. Thus Agni is of prime importance in the evolutionary process, and so we see in the Vedic texts such as the Rig Veda that Agni is indeed the first of the Devas to be invoked, with Agni being the very first word of the Rig Veda.

We shall see in the chapter on Devas that the Rig Veda primarily concerns itself with the Deva kingdom. This is because it is essentially a metaphorical and intuitive text, aiming to inspire or evolve the more logical or emotional human mind to the Deva kingdom of higher intelligence. More concrete (but still metaphorical) Vedic texts describing history and cosmology are the encyclopaedic Puranas (old texts), and the Itihasas (historical texts) based around epics.

Itihasa (History)

The two most famous Itihasas are the Ramayana and Mahabharata relating to the Devas Rama and Krishna, respectively. It is possible to trace the cosmology described in this chapter in these texts also.

The Ramayana, to summarise very briefly, is the story of Lord Rama whose wife Sita was abducted by the evil ten headed Ravana. This was during Rama's 14 year exile, enforced by his step-mother, so that her son could be king instead of the rightful heir Rama. During the exile, Rama was accompanied by his brother Lakshmana, in addition to his wife Sita. Rama was aided in successfully rescuing Sita from Ravana by an army of monkeys along with their commander, the monkey Deva Hanuman, who was a devotee of Rama, and who seemed to contain the miraculous force and power of the wind. In order to rescue Sita, waters had to be bridged to Lanka, the land of Ravana.

Here the cosmology is metaphorically represented by Sita as Mother Nature, or Prakriti, the link between Purusha as Rama and the world manifest in time. Sita corresponds to the bliss kosha. Lakshmana is the pure mind corresponding to the intelligence kosha. The waters to be bridged represent the sensory or emotional mind. Hanuman is the life-force or Prana, and the army of the monkeys represent the final food kosha, or the physical body. Additionally, the ten headed Ravana represents the five sense organs and the five motor organs, and Lanka his sensory world. The action of Rama's step-mother represents the factor of cosmology lying outside time.

Thus instead of Prakriti (Sita) achieving a stable and real unity as Purusha (Rama), downward motion to a denser and less evolved stability occurs due to the dirty fuels provided by the misuse of the senses (Ravana). As a consequence, the ahamkara functions as the lower self in the sensory world (Lanka). Hanuman, as the life-force and devotee of Rama, gives the urge to the state of real stability and unity of Rama. He manages to cross the waters (emotions) to Lanka (sensory world), and using the clean fuels of the monkey army (purified physical body), rescues or uplifts Sita to her position as transcendental Nature or Prakriti. Rama is the underlying reality behind this cosmology, and Lakshmana the guiding principles and cosmic intelligence.

The cosmic triad in the above is Rama, Sita and Hanuman (Purusha, Prakriti and Prana), but can also be Rama, Sita and Lakshmana (Purusha, Prakriti and

Lakshmana as the cosmic Prana in the form of intelligence).

Perhaps the best known of the Vedic texts, and the most important part of the Mahabharata, is the Bhagavad Gita. It is a conversation between Krishna as the cosmic person or Ishvara, and his disciple Arjuna. It occurs during a war between two branches of a family, the Pandavas and Kauravas. The Pandavas are lead by five brothers including Arjuna, and the forces of the Kauravas in the main consist of a 100 brothers. During the conversation Arjuna is in a chariot which has five horses, and his charioteer is Krishna.

Here Arjuna is part of a pentuple (five brothers) representing a higher level of the human kingdom, and he specifically represents (by his name) an aspect of the principle of Agni. He is trying to access the level of Ishvara or Krishna within himself so that he can use cosmic intelligence in order to control his senses. This is symbolised by Krishna being the charioteer of the five horses. Arjuna is fighting this war in order to provide the clean fuel of evolution and defeat the dirty fuel of downward movement. The dirty fuel is symbolised by the 100 Kaurava brothers as the ten sense and motor organs, which get afflicted in ten different ways each.[11]

ENDNOTES

1. The Sankhya Karika of Ishvara Krishna is the main text dealing with this.
2. Chandogya Upanishad 6.3-6
3. In the commentary of Swami Shivananda on Brahma Sutras 2.1.24-25
4. Note the works of Jaidev Singh on Kashmiri Shaivism and Abhinavagupta on this.
5. Yajnavalkya comments on this in the Brihadaranyaka Upanishad 3.8.4-7.
6. For example see The Self Aware Universe by Amit Goswami.
7. We see an example of this in the Chandogya Upanishad 3.13
8. A reference to the five elements in the Taittiriya Upanishad 2.1
9. The five koshas in the Taittiriya Upanishad 2.1-5
10. An example of the septuple in the Chandogya Upanishad 2.8
11. For much more on this see God Talks With Arjuna by Yogananda Paramahansa

3.

Mantra: Thought, Sound and Language

Om, This Syllable should be used for Upliftment.
Chandogya Upanishad 1.1.1

Mantra as Yoga

The role of sound is given much importance as an evolutionary tool, or a path of Yoga (Mantra Yoga), in Vedic thought. Evolutionary sounds are called mantras in Sanskrit, and mantra can be translated as an instrument of the mind. The reasons for the importance of sound follow on naturally from the preceding chapters. We saw that the space element is the least dense or physical of the elements, and that it corresponds to thought, sound and words. As a result it is the element that relates the most to the mental world, and so it is the closest to the pure consciousness or Purusha. In fact, the highest form of sound is ultimately the word that is the equivalent of the thought or will (Prana) of Purusha, and this sound is responsible for all manifestation, including that of Prakasha.

Thus it is not only physical sound that is important, but also the deeper level such as the tanmatra of sound, and the causal equivalent of the sound element at the level of the bliss kosha. At this level sound has a pure evolutionary power due to its close connection with reality and Purusha, compared to the lower levels of sound, especially physical sound. Causal sound or thought is a state of concentration and meditation, and it represents the birth place of language and ideas, or the actions and results of Mahat. So when we consider mantras, we do so at several levels, from the physical to the subtle.

Causal Mantras and Manifestation

Transcendental light as Prakasha is the basis of the energy of manifestation, and when Prakasha takes on in its more energetic form in order to facilitate manifestation, it is known as Shakti. The first form of manifestation that Shakti facilitates is the subtle or causal form of sound, or the sound of pure consciousness. It is the basis of creation as a projection of consciousness via word and language, or mantras.[1] This level of causal or creative sound is eternally present as it is rooted at the level of the onset of time itself, and it connects to the higher part of the Deva kingdom, as we shall see in the next chapter. Thus the sound of the Devas is responsible for all manifestation, including the tanmatras and the elements. In addition, the evolution of the ahamkaras through the kingdoms of nature via their manifestation of higher and higher forms is also dependent on the evolutionary power of causal sounds of the Devas. The ahamkaras have their foundation in reality at the causal level of the bliss kosha, which is the kosha of causal sound and the Devas.

The Importance of the Oral Tradition

From the above we can see how Mantra Yoga alone can give access to a transcendental state. The Vedic tradition holds that the Rig and the other Vedas are directly such an emanation of Shakti as sound, or mantras. The Vedas first occur at the causal level, and then at the mental level, and finally at the audible or physical level. In this way they serve to bring out the forces of evolution in manifestation. It is for these reasons that in the Vedic tradition there was so little emphasis put on the physically written word relative to the mental word.

To add to this, the relative proximity of the mind to the causal world of the bliss kosha gives the mental transmission of knowledge the ability to directly transfer the creativity, knowledge and contentment of the causal level, whereas the written word is more removed from it. The less we receive such real knowledge and contentment, the more our learning and experiences become unreal and hollow, and we feel unfulfilled at a deep level. Therefore, despite being composed in the very sophisticated linguistics, grammar and poetry of the Vedic Sanskrit language, the Vedas did not appear in written form for several millennia after their conception. This corresponded to a decline in the general ability to learn at the time, as will be explained in the chapter on Astrology. The purpose of the various Vedic sciences as outlined in the second part of the book is to help us overcome such difficulties.

Audible Sound and Mantras

The main evolutionary power of sound is at its most subtle level, but even the audible manifestation of sound has power. The principles of eternal intelligence used in manifestation are equally true regardless of whether it is at the gross physical level, or at the very subtle and causal level. This in itself is an important and directly perceivable eternal principle made use of in Vedic thought, i.e. principles themselves are invariant, just the density or quality of their manifestation that varies according to the level or kosha of manifestation. For example, the intelligent principle that peace requires stillness is true across all the koshas. We have physical peace via physical stillness, and mental peace via mental stillness, but they connect to and aid each other. The quality of peace varies according to the kosha the stillness exists at, with the higher koshas providing a greater level of peace, i.e. peace of mind is considered to be superior to peace of body. Another example is the rational principles of mathematics, which are not just true at a mental level, but are also true at the level of physical matter.

So we can see even audible sound has the intelligence of the higher levels of sound, and thereby it can connect us to them. However, it is just an entry point and has the least power of all the levels of sound. How successfully it can connect us to the higher levels depends on how natural the particular audible sound is, i.e. whether it has the evolutionary power of a mantra. It relates to the concept of a clean or a dirty fuel in terms of sound, and clean or natural sounds will be examined later in the chapter.

Three Levels of Sound

The various levels of sound and their evolutionary powers are commonly classified into three main categories, according to the system of the Yogi scholar of language and sound, Bhartrihari.[2] First there is audible sound known as vaikhari (varied) in Sanskrit. It relates to the food and Prana koshas due to its physical make-up via the air element, and so its influence is also strongest at these levels. The next category is called madhyama (middle) as it is a link stage between audible sound and the causal forms of sound. It is usually associated with whispered speech which has a similar linking property, and it relates to the Prana and sensory or emotional koshas where it represents thoughts as emotions.

The third category of sound is known as pashyanti ('seeing' or directly perceived), and this relates to the higher levels or koshas, or to the first emanation of Shakti as described. Here the speech of sound is like an unstruck and waveless sound made by the will of the mind, instead of by waves of air in the case of vaikhari, or by waves of the mind in the case of madhyama. Pashyanti works with the eternal intelligent laws and principles inherent in nature, and it is the equivalent of one-pointed concentration. Concentration on one thing by definition naturally brings us closer to a state of unity or reality, and this is a subject of the chapter on Yoga. These positive, uplifting and one-pointed thoughts of the causal levels are the seeds of evolutionary manifestation, and relate to what are known as bija (seed) mantras. In the next sections the basis of some of these mantras will be examined, including how the audible bija mantras connect to their higher counterparts.

There is also a fourth category in this scheme, but that relates to the sound in purely transcendental form as the sound made by the will of Purusha. It therefore goes beyond the concept of sound and no sound, and is known as para (beyond). Great Mantra Yogis such as Abhinavagupta[3] have given more categories of sound, but these essentially give more specificity to the above categories, especially at the higher levels.

Bija Mantras and Natural Sound

Bija mantras are particularly powerful and useful tools in the process of evolution towards real stability, or Purusha. However, to access these mantras in a deliberate manner at a causal level requires levels of stillness and concentration which are very difficult to attain. Therefore in Mantra Yoga audible equivalents are derived, and these can give access to the higher aspects of the bija mantras as they are based on the same intelligent principles. It is due to the principle that principles themselves are invariant across the different levels of nature, it is only their power or quality (or degree of manifestation) that varies according to the level of nature. We will attempt to demonstrate how this is achieved in the case of mantras by examining the natural basis of audible human sound and language. Causal mantras are naturally formed, so if we can derive a natural basis for audible human sound, we can arrive at a connection from the audible to the causal.[4] The symbols and metaphors of poetry and art similarly connect the inspirational levels of nature with its physical levels.

The first and most obvious place to examine in order to derive natural and audible human sound is the human mouth. The sounds that can be most naturally and simply made with the mouth should have particular significance. We can see that the first of these types of sound is to make a general 'aaa...' or 'uhhh...' sound (like the 'u' in the English word hum), with an open mouth, and then to let the mouth close naturally. It is the most simple and undisturbed sound we can make, and it encompasses the entire range of the mouth commencing at the back of the throat and ending at the lips. This, of course, is the sound Om or Aum. Hence at the higher levels Om is considered the first, most natural, complete and pure bija mantra, and it relates to the unified Deva Ishvara. As Ishvara contains all the Devas, so Om is considered to contain all the mantras at all levels, and this reflects the fact that it contains the entire range of sound that we can make.

From the natural starting position of the open mouth, there is only one other simple audible sound that can be made, for reasons of the structure of the mouth. Instead of closing the mouth effortlessly, the only other simple or natural way of closing the mouth is to widen it. This leads to the sound 'aaeemm' (rhyming with the English word lime), or the bija mantra Aim. By experimentation, we can see that there are no other variants of sound possible from this starting position of the mouth. Om and Aim thus are the two main bija mantras that are possible. As Aim involves some effort and energy in the widening process of the mouth, it relates more to primary energy or primary manifestation. Om and Aim therefore form a complementary dual like Purusha and Prakriti.

The Triad and Pentuple in Audible Sound

Notice that the mantra Om contains a 'u' or 'ooo' sound in the middle, and so it is spelled as Aum as well. In these two main bija mantras we have, in fact, three vowel sounds: 'a' and 'u' in the case of Om, and 'a' and 'e' in the case of Aim, giving the total of three distinct vowels: 'a', 'e' and 'u'. The m sound at the end of both of the mantras is only the closure of the audible sound of the main mantra. The importance of the closure lies in the fact that it links the audible part of the mantra to the inaudible part, which leads to mental concentration on the mantra. This takes us to a higher kosha of activity, or to the more subtle part of the mantra, and it is an important part of the practice of Yoga.

The u sound in Om relates to purification or a clean fuel, i.e. an uplifting energy (it relates to the word up in English). Its pronunciation causes the mouth to take a more vertical or up and down shape, and so Om also has the capacity or force to uplift. The three vowels 'a', 'e' and 'u' form the vowel sound equivalent of the triad of the Sankhya system. These vowels are the only three simple audible vowels that can exist by the mechanics of the mouth. Thus we have (note that these definitions cannot be rigid, especially at the higher levels):

a: the first possible sound, and so relating to our primary sense of existence (Purusha)
e: projective or energetic in nature (Prakriti)
u: purifying or uplifting (Prana)

Due to the primacy of these vowels, we have natural instincts as a human race to use these sounds, as in the expressions 'ahh', 'ooo' or 'eee'. On reflection, we can observe the above type of experiences when doing so.

If we include the motion of the tongue in addition to the mouth, two more natural vowel sounds are possible. The tongue naturally requires the palate in order to make a sound, and this can be done in two ways. The first is with motion as in the 'ṛrr' sound, and the second is without motion, as in the 'lll' sound. As these sounds can be held in time they are classified as vowels, and this brings the total number of natural vowels sounds that are possible to five. The 'r' sound is very energetic in nature, and so it corresponds to energy and transformation, or to a fiery nature. The 'l' sound is more supportive (the tongue supports the palate while it is being made), and thus it corresponds to a nourishing type of quality.

Sanskrit, the Natural Language of Mantra

These five natural vowel sounds that are possible form the basis of the vowels of the Sanskrit language. There are other vowels in Sanskrit such as the vowel sound in the words lay and lie, as well as the vowels sounds in the words low and loud, however they are of secondary importance as they are combinations of the primary five vowels. In addition to the vowels, there are five groups of consonants, each containing five consonants. Each group corresponds to the five different regions of the mouth that can be used for making sounds, i.e. the throat, back of the palate, front of the palate, the teeth and the lips. In each of these regions there are five variations possible, which gives rise to the five consonants of each group. For example, the pentuple of consonants corresponding to the first group of the throat are:

(The 'a' after each consonant is pronounced as the 'u' as in the English word hum. It is there in order to make the consonant pronounceable, as a consonant cannot be pronounced without an accompanying vowel.)

1.ka 2.kha 3.ga 4.gha 5.gna

The last consonant in the pentuple is a type of nasalisation representing the closure of the series, a bit like the 'm' in Om, which is also akin to nasalisation. The consonants with the 'h' are the aspirated versions of the preceding consonant, or versions that are energised with the extra use of the breath. From observation we can see that these are the only possibilities for consonants in this group. Each of the other groups follows a similar line of reasoning. It is useful to get an insight into the structure of such letters, as then we can see how bija mantras are constructed, as well as what type meaning and effect they can have. In addition to the vowels and consonants there are the semi-vowels (including sibilants) such as 'sha', 'ya' and 'va'. There is one pentuple and one triad of these, corresponding naturally to the different regions of the mouth.[5]

From the above analysis we can see how Sanskrit is based on the language of nature, and thus how it is the language of bija and other mantras.

Some Important Bija Mantras

In addition to the two main bija mantras of Om and Aim, we can now derive some other important bija mantras, and we can also get an insight into their higher meanings at the causal level. First we should take a look at the qualities of the triad of consonants, vowels and semi-vowels generally.

Consonant sounds relate more to concreteness as they are discrete, and vowel sounds relate more to permanence as they are steady and open-ended. Consonant sounds therefore are more oriented to the outer levels of manifestation, and vowels sounds are more oriented to the causal levels. The semi-vowels share properties of consonants and vowels, and so they act as the link between the two. Hence they relate more to the mental levels in-between the causal level and the outer levels of manifestation.

In relation to the consonants, 'ka' is the first of the consonant sounds that we can make, because it is the first sound that is possible from the back of the throat, which is the source of audible speech in the body. This is analogous to the first form of manifestation, and when we add the energetic 'ra' sound to it as 'kra', we derive the bija mantra for energy of primal manifestation or action. 'Kra' is thus the root for words relating to action in Sanskrit, including for the word karma (doing). When we add the sound 'eee' to this as 'kreee', the meaning of this combination is an energy ('ra') that changes primary or old forms ('ka'), to new ('eee') forms. With the closure of the nasalisation, the bija mantra becomes 'kreeem', or more commonly spelt as kreem.[6] It is considered a fierce bija mantra as it forces very energetic (due to the 'ra' and 'eee' sounds) changes in levels of stability, and as such, it can cause instability. Its evolutionary power (and attraction) lies in the fact that it can facilitate rapid movement to new and higher levels of equilibriums. However, this is dependent upon the type of fuel it has to work with. When the energy for change burns on a clean fuel (a fuel that that takes us towards the peace of reality), the mantra is beneficial and less fierce. A dirty fuel will cause the mantra to make us experience the instability and pain of forced change, without taking us towards reality. Hence this mantra is best performed after appropriate purificatory practices, and with much caution. Purification will be looked at in the chapters on Yoga and Yajna.

The bija mantra kleem is derived similarly to kreem, with the energetic 'ra' sound being replaced by the supportive 'la' sound. Therefore it relates to the gentler and more pleasant changes of forms. The bija mantra shreem has very supportive properties as it makes use of the 'sha' sound which has supportive and calming properties for similar reasons as 'la', but even more so due to the broader contact of the tongue on the palate. It is also energised by the 'ra' sound, which greatly increases the energy of its supportive nature. As there is no 'ka' in shreem, it relates to the concept of support generally across all levels, rather than at the onset of activity or manifestation. The bija mantra hum (pronounced hoom) has the uplifting and purifying power of the vowel 'u', as well as the energy of 'ha', which has a deeper but less vigorous form of energy than 'ra'. As a result, hum is an energetic and purifying bija mantra. The bija mantra hreem has the deeper (relating to breath and 'ha') as well as the vigorous ('ra') forms of energy. The vowel 'e' gives it a projectory power, and thus overall it corresponds to the source of energy of manifestation. This is much like the heart, and it helps us to go to the heart of things.

An important point to beware of is that the practice of

powerful mantras such as bija mantras should be done with the help of an experienced and knowledgeable guide, especially for newcomers to the area. These mantras can have strong psychological and physiological effects as has been discussed, and though these are usually beneficial, sometimes when they are done in an inharmonious manner they can become the equivalent of dirty fuels and can have negative effects.

The Vedas and Mantra

The language of the Vedas itself is said to be the language of mantra, and it connects us to the levels beyond the human kingdom. We noted that the Rishis are a level of beings above the human level in the Vedic system, and they connect the human kingdom to the Deva kingdom. In this linking role, the Rishis facilitate the connection between the causal sound of the Devas and the lower kingdoms via the Vedic mantras. In this manner the Rishis are able to aid the evolution of the human kingdom as its guides. The language of the Vedas, Vedic Sanskrit, is therefore a highly metaphorical language. Unlike the bija mantras, whose main focus is to try and connect us to the highest causal levels, the metaphorical language of the Vedas allows them to be considered at various levels, i.e. physical, metaphorical or intuitive, as well as causal. This gives them a much broader usage, and the Vedas are therefore a linking text, much like the Rishis and their purpose themselves.

We have seen in this chapter that there is a natural connection in terms of manifestation and evolution between the Devas and Rishis, as well as the Devas and mantras. These connections will be looked at in the next chapter on the Devas.

ENDNOTES

1. Bhartrihari Vakyapadiyam (and Commentary) covers this extensively, though it is a technical text.
2. As above.
3. See Tantrasara and Tantraloka of Abhinavagupta. It is currently available in Hindi and Sanskrit only.
4. As above.
5. The publications of Samskrta Bharati are a good resource in order to learn Sanskrit.
6. For more information on bija mantras refer to Mantra Yoga and Primal Sound by David Frawley.

4.

Devas: The Light of Manifestation

While Manifesting the Light of Reality, the Devas said that one who knows Reality will control them.
Yajur Veda, Taittiriya Aranyaka 3.13.5 (Purusha Suktam)

Devas in Manifestation

The Deva kingdom can be thought to consist of a higher part and a lower part, as we have seen. The higher part relates to the causal aspect of all manifestation, including time, and it is the part commonly associated with the Devas. The lower part lies above humanity in the evolutionary scale, but it occasionally does manifest in humanity as its guides, or Rishis. It is the link between the human kingdom and the higher part of the Deva kingdom, with the Rishis helping to shape the course of humanity by giving it spiritual leadership. In this chapter we will mainly look at the first part of the Deva kingdom, but we will also look at the second part, particularly in connection with the writings of the Rishis (The Vedas) in relation to the first part of the Deva kingdom. (For the purposes of this book the first part of the kingdom will be referred to as the Devas, and the second part as Rishis or Yogis.)

As the Deva kingdom occurs with the onset of time, it is eternal in time. This is in contrast to Purusha which lies outside of time altogether. The concept of time does not exist for Purusha, whereas for the Devas the concept does exist, but they abide for all of time. The Devas must actually be only one as mentioned earlier, as there cannot be qualities or differences between them due to their eternal nature. In the Vedic system, the Deva kingdom is presented as consisting of many Devas in order to illustrate its infinitely varied role in manifestation. It is similar to the paradox which we saw with Vedanta, i.e. how the one transcendent reality can appear as the infinite variety of nature. Thus the Devas can be seen as infinite in number too, but generally broader categories are made such as the 33 main Devas of the Rig Veda mentioned later in this chapter.

The Devas are the light and energy of Prakasha, and so they are the manifestation, and the means of manifestation, of transcendental consciousness, intelligence, Prana and time. Therefore they are actually inherent in the structure of all manifestation, including physical manifestation. As facilitators of the cosmic process, they are also the means how manifestation evolves towards the reality of the causal world, along with the aid of the Rishis.

Devis – Kali and the Establishment of Time

As the Devas are responsible for guiding or nurturing manifestation and evolution, they are considered to form the womb of manifestation. The is the golden womb of all manifestation with its 'cosmic waters', and it is known as hiranyagarbha in Sanskrit. The Brahmanda, or the cosmic egg from which the universe is born, is incubated in it. As such, it is the female aspect of the Devas, or the Devis, that is of prime importance. The Devis are the nourishing energy of manifestation, or Shakti, in contrast to the pure light form of Prakasha. Much as Prakriti establishes the duality with Purusha necessary for manifestation, the Devis form a similar duality with the Devas, and in their unified forms the Devas and Devis are known as Ishvara and Shakti. Thus from the perspective of manifestation, which is the perspective of our thoughts, it is the Devis that direct us on our evolutionary journey first and foremost. The first function the Devis perform in order to nurture manifestation as the energetic form of light is to establish themselves as time, since the causation of manifestation requires time. (We can note the principle of the dependency of time on light from earlier, and as even recognised by modern science.) Therefore the first of the Devis is Kali, the Devi of time, by whom the variety of Prakriti is able to manifest. She creates time by her causal thought or mantra (the relationship between Devis and mantras is described in the next section). This thought forms the basis of our memories and imagination, i.e. the past and the future, and thereby it establishes time.

Devas, Devis and Mantras

In the chapter on Vedanta we saw that that the only reality (Brahman) is a complete consciousness and intelligence. Via the will of this consciousness, manifestation is visualised as the dream of Brahman. In order to facilitate this dream, time along with a primary level of thoughts that occur with time, are projected from the consciousness of Brahman. These levels of thoughts equate with the causal mantras, as well as the equivalent concept of the Devas and Devis. Further levels of manifestation and evolution are then the dreams or projections of the Devas and mantras. Dense physical manifestation can be understood simply as thought objects of the Devas as their mantras.

This is not as remote from us as it may seem, but it must be actually happening here and now, including within us. It is the eternal process of manifestation that constantly happens now, for there is only ever a now. Our conscious awareness of our existence is Brahman. Our awareness of the external world is the projection of our Deva consciousness. Our sense of progress and evolution is via our Devas of the intelligent principles of balance and stability which draw us towards Brahman consciousness. Similarly, our sense of separation, aggression and depression occurs when we let our Deva consciousness of the process of manifestation overwhelm our Deva consciousness of the process of evolution. We confuse a lower stability for a higher one, and this is fundamentally depressing, as it takes us away from real stability. This entire play simply represents our state of mind at various levels, from the state of mind of an individual, to the collective state of mind of a kingdom of nature, to the universal state of mind.

A particular example of how the Devas and mantras relate to each other is when we consider the manifestation of the cosmic triad of Purusha, Prakriti and Prana/Prakasha (combining primary light energy with its force) into time. It is reflected in the first division of Ishvara into three as Shiva, Vishnu and Brahma, an important part of the current Hindu tradition, and who also relate to the triad of the Father, Son and Holy Ghost in Christianity. We could say Purusha relates to Shiva, as Shiva is known for the destruction of manifestation, leaving only the reality of Purusha. Vishnu is the preserver or the basis of manifestation, and so relates to Prakriti. Brahma is the light or energy by which manifestation takes place, and so relates to Prakasha. These Devas then become the foundational building blocks and guiding intelligences of all manifestation and evolution, as a triad. As these three Devas are in reality one, there is much overlap between the three, and so we should realise that a strict rational compartmentalisation is not appropriate.

The rest of manifestation below the triad of Shiva, Vishnu and Brahma is the projection or dream of their consciousness. This requires Shakti or energy, which by definition must be supportive, i.e. must be a Devi energy or projection. Therefore each part of this triad has a female counterpart, with Shiva relating to Kali, Vishnu to Lakshmi and Brahma to Sarasvati. The qualities of Devis are the same as the Devas, with the Devas acting more as a potential or force, and the Devis acting more as energy and action. As this energy of manifestation occurs at a causal level, the Devis especially can be correlated directly to causal mantras. For example, we can correlate the audible bija mantras from the previous chapter to them.

Kali constitutes the energy of creative destruction being the female counterpart of Shiva. This is a role of time, as all manifestation begins and comes to its end due to time. Kali represents this type of cyclical energy of time, and so she connects to the bija mantra kreem which creates in time, but at the same time its fierceness destroys old forms. Lakshmi is supportive in nature, and so she connects to the bija mantra shreem. The energy of Sarasvati relates to general manifestation, and so she connects to the bija mantra Aim.

We should note that as Ishvara underlies the triad of Devas and Devis, it relates to the mantra Om, which similarly underlies all other mantras as the most natural mantra.

The Devas of the Rig Veda and 3, 5 and 33

The Rig Veda is a vast source of mantras relating to the Devas, with all its verses connecting to particular Devas or Devis. This is so as it is an inspirational text which aims to uplift or evolve us to a higher state of consciousness, or to the plane of the Devas. Its mantras are the visualisations of the Rishis or seers of the higher part of the Deva kingdom. It relates to many Devas in order to reflect the divinity in the diversity of cosmology and manifestation. To this extent it reflects the principle of the triad and the pentuple. Three of the main Devas in it are Agni (cosmic fire), Vayu (air) and Soma (cosmic water), which form a complement, and which reflect the guiding principles behind all manifestation, as discussed in the chapter on Cosmology. The pentuple of Devas is formed in the Rig Veda with the inclusion of two of the other main Devas, Indra and Surya. Indra represents the divine mind or divine ego, and Surya represents the divine intelligence.

These five major Devas of the Rig Veda also represent the five koshas, due to the fact they form a pentuple. As Devas are the basis of all levels of manifestation, any of these five Devas can represent any of the koshas. However, in this case Indra is more commonly regarded as the first level of individualisation, or the jivatman associated with the bliss kosha. He is portrayed as the main drinker of Soma, the nectar of bliss. Surya relates to the intelligence kosha. Agni, Vayu and Soma each can represent any of the three lower koshas in their forms as the Ayurvedic doshas, as we shall see in the chapter on Ayurveda. However, there is a natural connection with Prana and Vayu as the symbol of air, as air like Prana, reflects the potential or force of all motion.

The role of Agni was particularly highlighted as a key driver or energy of the evolutionary process. Hence he is the first and most frequently occurring Deva in the Rig Veda, and so it is possible to view all manifestation and evolution as the movement of Agni. For example, we have five levels of fire corresponding to the five elements, or any of the pentuples.[1] At the level of the earth element, Agni corresponds to the energy in matter, which is the level of the food kosha. It is the inert fire present in matter which is released when we burn substances like wood. At the level of the water element, Agni relates to the biological fire of the instinctual kosha. Here it is known as Pitta dosha (as we shall see in the chapter on Ayurveda), for which fuel is symbolised by water. At the level of the fire element itself, it corresponds to the physical fire we normally experience. It relates to the sensory kosha as it activates our senses such as sight. At the level of the air element, Agni corresponds to atmospheric fire which relates to the kosha of intelligence. Here the mind behaves much like air in the way it moves. At the level of the space element, Agni corresponds to spatial fire which is solar or stellar fire. This is a complete fire as it is all pervading, and thus relates to the kosha of contentment or bliss. It is also the source of all fire, much like our own source, the jivatman associated with the bliss kosha.

The main group of Devas in the Rig Veda is the extension of the triad from 3 to 33. It reflects the variety that exists in manifestation and cosmology for which the Devas are responsible. This group of 33 breaks down into a pentuple and a triad. The triad is represented by the Agni, Vayu and Soma groups of Devas, broadly speaking. The Agni or energy group comprises of the 12 Adityas, or the Devas specifically relating to cosmic or stellar light. They correspond to the division of the stellar light into 12 in the form of the zodiac or the months. The Vayu or Prana group comprises of the 11 Rudras, and they relate to the action of Prana in the five sense organs, five motor organs and the mind. The Soma group consists of the eight Vasus who relate to the basic and universal substances of manifestation. These eight substances are a naturally ascending order of the way our world presents itself to us, from is lowest depths to its furthest reaches. They are:

1. Fire (at the core of the earth)
2. The Earth
3. Air
4. Sky (Clouds)
5. Moon
6. Sun
7. Heaven (Astronomical Space)
8. Stars

In addition to these three groups with a total of 31 Devas, there is Indra as the divine mind or person, and Prajapati as the divine intelligence. This makes up the total of 33 Devas. Along with Agni, Vayu and Soma groups, Indra and Prajapati also make up a pentuple.

In the Rig Veda, all of the Devas, including the 33 above, are said to be born of the golden womb of the great Devi Aditi. Thus Aditi relates to Prakriti and Maya of the later Sankhya and Vedanta systems, respectively. Maya is the great mother of all that appears to be real whether earthly or divine, but is actually illusory as explained in the chapter on Vedanta. So yet she is not the mother of what is real, or Brahman.

Devas, Humans and Evil

We saw that the Devas are responsible for manifestation and evolution from the causal level. We saw further that nature across all its levels can be considered as a manifestation of the different levels of Prana of the Devas themselves, via the energy of Shakti. The causal Deva kingdom is where the natural intelligence of light occurs in its purest form. In the human kingdom, the intelligence of the Deva kingdom manifests in a lower form, and the clarity of thought required for causal activity is also much lower. Directly perceived truths begin to be dominated by logical thinking and emotions. As a result, creative original thoughts such as the great new ideas of humanity, which relate to the creativity of the causal level, are sporadic. (Logic and deduction alone cannot generate anything new, but can merely rearrange existing information.) In the Vedic system, the Rishis separate themselves as a kingdom from the humans due to their consistently great insights into philosophy, language, mathematics and arts from ancient times (the subjects of the Vedic texts).

Due to the domination of logic, emotions and senses at the expense of the unity of directly perceived truth, the decision making faculty of the human kingdom becomes confused at times. This leads to the ups and down in the paths of our evolution towards reality. The senses and emotions create variety, and naturally separate us from the unity of reality if they dominate us. They then become dirty fuels, which results in the downward motion of a sense of separation, rather than the upward motion of evolution. This naturally increases emotions such as depression, fear and aggression in us, as it takes us away from the peace and stability of Purusha.

The above type of psychology is what is known as the world of the demons and 'evil', or the archetypes represented by figures such as Ravana and the Kauravas described in the Itihasa section earlier, or Satan in the Christian tradition. The Devas appear to take on an 'evil' form in this aspect of the human kingdom, but the concept of 'evil' is relative, as it is a necessary stage in the evolutionary process. At each stage of evolution, the weakness of that stage and the strength of the next have to be experienced in order to maintain the momentum of the evolutionary process. Hence progress always must have a certain mix of ups and downs. It is a reflection of the fact that reality beyond dualities underlies, both, 'good' and 'evil'.

Due to the fact the decision making faculty is reduced from the level of logic and emotions to the level of instincts in the kingdoms of nature below the human kingdom, the ups and downs of evolution are much reduced there. The decision making faculty is almost non-existent at these lower levels. Therefore the concept of 'evil' is particularly centred around the humans, but on the other hand, the ups and downs of evolution also bring a faster rate of evolutionary progress.

We should note that the kingdoms of nature are not discrete, but merge into each other at their margins. Hence the lowest of the Rishi class may manifest some negative traits like humans, and the highest of the humans can manifest qualities of the Rishis.

The Role of Metaphors

As the Devas enable the cosmic triad to become the physical triad, the lower kingdoms of manifestation are in fact just lower forms of the Devas. It is an example of how principles, such as the principle behind the Devas, remain invariant through the kingdoms of nature. This is an important point in the evolutionary process as it indicates how the lower kingdoms of nature can help with the evolution of the higher kingdoms. It is the basis of the use of metaphors in the Vedic tradition, and to this extent, the Devas are known via natural metaphors for their divine qualities. In addition to the metaphors of fire, air and water for Agni, Vayu and Soma that we have already seen, further examples of nature's metaphors will be described in the next section.

The importance of the role of metaphors is further demonstrated by the dynamic interplay between the various levels of manifestation. As concepts such as time and intelligence make up manifestation at all levels, it is possible for the various principles, kingdoms, elements and other aspects of nature to merge into each other in an inter-woven and non-linear fashion. Evidence for this has also emerged from modern science, where Relativity Theory and Quantum Mechanics have a requirement for exactly this inter-mingling and non-linearity of matter, energy, space and time. The discrete system of classification and categorisation is one of the limitations of rational thought. Rationality is one of the main tools of humanity, and as such is very powerful, but requires complementary approaches too.

Therefore we see why metaphors are so central to the Vedic tradition, especially the Rig Veda. Metaphors appeal to thoughts that lie beyond rationality, and in doing so, they aid direct perception. When discussing transcendental concepts such as the Devas at the onset of time itself, the use of metaphor becomes all the more important. To illustrate these points, in the Brihadaranyaka Upanishad Yajnavalkya uses the metaphor that everything (including the past, present and the future) is woven together, warp and woof, in unmanifest space (space at a causal level).[2]

Nature's Metaphors

A particularly important source of metaphors is nature itself. A result of the lesser decision making ability of the lower kingdoms of nature is that they are able to reflect the evolutionary process of the Devas in themselves in a purer way, albeit at a lower level than the humans. This is significant, because it makes the lower kingdoms the purest access to the purpose of the Devas available to us, as most of us do not have sufficiently developed perception to access the higher levels of truth of the Devas directly. The animal kingdom becomes particularly important in this regard, because it is the most advanced of the lower kingdoms. It can act as a pure reflection or metaphor of divinity for us, if we so choose (as can all of nature). Thus we see that the Vedic literature is particularly strong in its metaphorical use of animals, as well as other aspects of nature.

The cow is one of the foremost of such metaphors[3] used. For example, it is widely used in the Rig Veda, and also in the story of Krishna. Traditionally the cow has been a source of sustenance for humanity via its milk and its work in agriculture. Similarly, in the Rig Veda the cow often refers to mental sustenance, or light and knowledge as a nourishing energy. Krishna, as a form of Vishnu, is the great sustainer and supporter. He is known as lord of the cows, or Govinda. In the Rig Veda there are references to the seven words (saptavanih)[4] as the seven streams of sound from the cow of mother nature (Prakriti). Here the septuple shows the complete range of sound (mantra) and light (Deva) as sustenance, from the transcendental to the physical (the septuple represents completeness). It is analogous to the seven visible rays of light in the spectrum, the seven separate notes in an octave, and also the seven days of the week (the day being a measurement of time via sun-light).

The horse is often used as a metaphor for the power or Prana aspect of nature, due to the obvious strength of the horse. The horse sacrifice[5] in the Brihadaranyaka Upanishad refers to the sacrifice of all the power in manifestation to Brahman, or that all power is of Brahman. The swan refers to the calmed mind of an advanced person or Yogi, much like the calm yet graceful and elegant movement of the swan. The eagle refers to great vision and mental acuity (the ability of the mind to soar to great heights), like the vision and the precise movement of an eagle.

The bodies of water on the earth are frequently used as metaphors too. The oceans and seas refer to the cosmic water which gives birth to all of manifestation, like how the sea is the source of all life on earth. The mountains refer to the higher states of consciousness, like how the air is rarefied at the heights to which the high mountains soar. The river is like a flow that links these higher states of consciousness to the lower physical planes, or like a source of inspiration. The great Himalayan rivers such as the Sarasvati (now more or less dry), Indus, Ganga and Brahmaputra are used extensively as such a metaphor. In Vedic times the Sarasvati was the foremost of these metaphorical rivers, and it appears that it was also the greatest of the Himalayan rivers. In the Vedas, Sarasvati is the main Devi or Shakti of knowledge, speech and inspiration.

From these examples we can see that we can learn something profound from all of nature, whether it is from the feel of a stone, the sound of birds or a stream, the fragrance or vision of a flower, the space of the sky, the trust or faith of an animal, or the stability of a tree.

Sanskrit, the Natural Language of Metaphor

It is no surprise to find that the metaphorical relationship between the different levels of manifestation is also reflected in Sanskrit, the natural language of mantra and the Vedas. Sanskrit is a highly metaphorical language, with the same word having a range of interpretations from the earthly to the divine, yet connected by a common underlying principle. This underlying connection is much like the River Sarasvati, the Devi of speech and knowledge. We have already seen many examples of this, such as with the word Agni, which can range in meaning from a physical fire to the cosmic fire. Often this use of the language has been lost in the more logical or modern approach to the Vedas, where a mechanical and materialistic interpretation to life is given primacy. As a result it is not necessarily correct to think of the Vedic and other ancient societies as undeveloped. They in fact had a deeper but less materialistic knowledge than modern society.

ENDNOTES

1. Chandogya Upanishad 4.11-13
2. Brihadaranyaka Upanishad 3.8.4
3. See number 4.
4. Secret of the Veda, Chapter 11
5. Brihadaranyaka Upanishad 1.1.1

Part 2. The Individual (Pindanda)

The Individual Self is the Universal Self.

Mandukya Upanishad 2

5.

Vijnana: Vedic Science

Knowledge is the One Reality.

Aitareya Upanishad 3.3

The first part of the book indicates why we need the concepts of the universal (Brahmanda) and the individual (Pindanda), whereas the second part gives us the tools of the Vedic sciences to re-establish the unity of the universal and the individual. In this way the Vedic sciences are the application of the Vedic philosophy and cosmology at the level of individual daily living, and they form the more practical second part of the book. This chapter on the overview of the Vedic sciences particularly serves to connect the first part of the book as the universal, with the second part as the individual.

Vedic Science and Modern Science Revisited

A question that often comes up in relation to the Vedic tradition is 'Of what use is Vedic science now that we have modern science?' Surely modern science has taken the concept of inquiry to new levels, especially when coupled with the technology it has at its disposal in order to perform its analysis. Perhaps Vedic science had its use in the past as a sort of 'proto' or primitive science, but with the progress of technology its role is no longer relevant, apart from historical and anthropological reasons.

However when we consider inquiry from the basis of a fundamental or a stable reality, we can see the purpose of Vedic science clearly, as was introduced in the first chapter.

We call something real at a fundamental level when it is permanent or stable. Things that keep changing are relative, so and their truth is limited to an interval of time, i.e. they cannot be described as an absolute truth. In order to describe something as real, rationality demands that its truth be absolute or permanent. Both modern science and Vedic science seek such a type of reality as an underlying basis of all phenomena, by their varying methods of inquiry. There are some limitations that modern science has imposed on itself in this regard, and this is an important way it is differentiated from Vedic science, and this is also why a Vedic approach to science is essential for any inquiry into reality.

One of these limitations is that modern science has not considered the nature of a permanent reality sufficiently, especially with regard to how it underlies any type of scientific inquiry, much like a subject underlies the observation of an object. Instead it has chosen to take the first available data that it has, and more or less by default consider that as its absolute basis of reality. This of-course is the data we have available to us via our senses. It is obvious even by the results of modern science that this is quite unsatisfactory, as sensory data has no stability at all. Each state of sensory data is in a constant state of flux, and only has an infinitesimal existence in time. This gives modern science no fundamental basis or foundation to stand upon.

This limitation is particularly highlighted when we consider the duality of the subject and the object, as was referred to in the first chapter. Modern science defines all existence to exclusively comprise of sensory data, and this includes both, the observer as subject, and the observed as object. However in order to make its observations, rationality requires the observer to be different from, or completely uninfluenced by, the object. Otherwise the observer cannot really make any fixed observations (which it seems we can), as the process of observing an object will change the object when the observer and the observed are the same. In this case the object of observation will no longer exist, and the observer will never be able to 'catch-up' with any thing they may wish to observe. In the field of logic this is referred to as a problem of self-reference or circularity. It is analogous to standing in front of a mirror and trying to observe what you look like when your eyes are shut. Here the subject and object are both the same, i.e. yourself. The act of trying to make this observation requires you to open your eyes, and this makes the object of your observations (yourself with your eyes shut) no longer exist.

The importance of the Vedic approach to science is that it does postulate an observer independent of any sensory data. This implies that the observer must have a non-material aspect, but the Vedic sciences go even further than that. According to them in order to make observations in time, there must be a conscious observer which is also independent of time due to reasons of self-reference again. Such transcendental observation is known as sakshi in Sanskrit, and it implies that the subject and object must eventually unite. It leads to the conclusion that manifestation can only be pseudo-real, as we have seen.

Ultimately there is actually no such thing as Vedic science or modern science, but there is only a rationally consistent science as a form of inquiry into reality. The Vedic approach to science is this type of science. Modern science appears to have become unsound at a fundamental level as we have seen, so at a deeper level it is a pseudo-science. In fact the great pioneering thinkers of modern science and philosophy (from recent times as well as earlier periods) also had a Vedic approach to science. These include thinkers (virtually all of whom are Nobel laureates, except naturally for those who pre-date the prize) such as Heisenberg, Schrodinger, Oppenheimer, Einstein, Schopenhauer, Kant, and even the great Greeks from earlier such as Plato.[1] It is at the wider level that the approach of modern science is fundamentally unsound, as the thoughts of these pioneering thinkers spread through the scientific communities.

That is not to say that modern science is of no use. When we consider sensory data without the perspective of a deeper inquiry into reality, it is exceptionally well developed. As a result it is naturally very useful in the routine of daily living. In today's society where there is a strong emphasis on materiality and consumerism, this is even more powerful. However if we ignore the underlying reality in our life and thinking, we feel a deep discontentment and depression as it contradicts the most fundamental and important principle that we know. This is an urge towards the natural stability of a unity that we can directly perceive. A complete science would thus need to consider the entire spectrum, from the sensory world to the underlying reality, as part of its inquiry into reality. It should build from the foundation of reality, and base its knowledge of the more superficial sensory world from this perspective. Modern science has reversed this approach, but as a positive it has developed its great material skills.

For these reasons the Vedic sciences prefer to concentrate on an inquiry into reality via direct perception, and give a lower priority to the sensory world. To this extent the Vedic sciences are based on the qualities and principles of nature itself, and so are eternal. Thus they do not change or evolve, but it is the human understanding of them that evolves over time. Due to this acknowledged importance of subjectivity and reality in the Vedic sciences, they are not simply sciences, but are also forms of music, poetry, philosophy, spirituality, and life itself. The clear conception of subject and object that is the basis of the Vedic sciences is also like the duality of Purusha and Prakriti (or science and art) upon which all manifestation rests, and this makes them particularly suitable for inquiring into manifestation and reality. It is what makes the various Vedic sciences actually overlapping facets of the one science of reality. The great Yogi Abhinavagupta was an embodiment of this type of eclectic thinking, and he showed the connection between aesthetics, art, philosophy and reality, in particular.[2]

Types of Knowledge

Another important difference between Vedic science and modern science is in the system of epistemics, or the theory of knowledge that they use. The Vedic sciences are based on the highly developed Vedic system of epistemics, which is a kind of knowledge of knowledge.[3] It is important for any science to have this, as it indicates the ways knowledge can be increased, and also the ways that it is limited. As such it increases the scope and efficiency of science, as well as giving a better understanding of its purpose. For example, it indicates the importance as well as the limitations of rationality,[4] an area of some confusion generally. Modern science by contrast has no such clear theory of knowledge, and this leads to a haphazard approach in its inquiry into reality, as well a confused idea of its purpose. We see today that modern science generally does appear to be losing its direction and purpose. In this section we will consider the roles and complementarity of rationality and direct perception as systems of knowledge.

Vedic epistemics begins by the division of knowledge into a subject and object, or as it is more commonly referred to in this case, the knower and the known. This also implies a third leg which connects these two, which is the process of knowing. The third leg is like the connection of Prana between Purusha and Prakriti at the cosmic level. When these three legs unite, we have pure knowledge or Purusha only. In Sanskrit the knower, the process of knowing and the known, are known as pramatri, pramana and prameya respectively. Each of the three legs above can be further divided into three or five as indicated below.

Pramana, or the appropriate means of gaining knowledge, is divided into three as two types of pratyaksha or perception, plus deduction. The first aspect of pratyaksha is sensory perception or what is commonly understood as induction, and the second aspect is the higher form of perception, or direct perception, as already mentioned. Deduction is obviously based on logic and rationality. The knower as the transcendental sakshi is divided into three as the Sat, Chit and Ananda aspects. The known is the field of manifestation, and here the triad expanded into the pentuple as we saw in the Sankhya system and its concept of the five elements.

The six systems of Vedic philosophy mentioned in the first chapter are also based on this system of Vedic epistemics, though there are some minor variations between them. For a further elaboration on Vedic epistemics, the works of Shankaracharya (perhaps the greatest Yogi and philosopher known) such as Drik Drishya Viveka are excellent and worthy of examination.

The Primacy of Direct Perception

Direct perception gives us awareness of our own being or self (as the subject or pramatri), as well as knowledge of areas such as language of mantra, logic, rationality, original or new ideas and ethics. It is also the inspiration behind art and poetic metaphor, and includes any knowledge we may gain which is not accessible via sensory data or the rationality of logic. It is the basis of the deeper aspects of the Vedic sciences and Vedic knowledge in general, by giving access to causal sound (mantra) and causal light (Deva). It is the difference between knowing something intellectually, or apart from oneself, and the practice of experiencing it directly where the subject and the object unite as one. As such, it is self-evident and complete in itself, and this alone can give the fulfilment of the complete knowledge required to transcend the finiteness of rationality and manifestation into the infiniteness of the universe and Purusha. Ultimately it is the perception or awareness of directly being reality only, or the perception of the sakshi, which transcends all mental activity and is the source of all knowledge.

Direct perception relates to the highest koshas of bliss and intelligence, whereas the sensory aspect of pratyaksha relates to the emotions and senses which is the area of the manas kosha. The last aspect of pramana is deduction, and this relates to the rational intellect which is a part of the intelligence kosha.

The Self-Referential or Incomplete Nature of Rationality

In contrast to direct perception, rationality requires a separate subject and object of knowledge, i.e. it requires the finiteness of manifestation. This limits it to intellectualism where knowledge is apart from the subject, as it is unable to unite with or comprehend infinite reality. Its finiteness is evident in the fact that it is not complete in itself, i.e. there is no rationality for rationality and it is eventually self-referential.[5] Thus rationality itself demands that it is limited, and it needs direct perception in order for the truth of existence, including its existence, to be realised.

The importance of rationality lies in the fact that it prepares the ground of every day life for the higher knowledge of direct perception. It helps to hold and organise any knowledge that is received via direct perception for the purposes of every day life, such as via the creativity and the inspiration in the arts and sciences. To this extent we should note that while rationality itself works within time and is finite, the principle of the truth of rationality is eternal or infinite as it is based on direct perception. It is why we naturally know that whatever is irrational is also unreal.

Knowledge and Faith

When we apply the concepts behind pramana to the controversial area of faith, it can help to bring some clarity. The faculty of direct perception is not usually as well developed in us compared to the faculty of logic, and thus generally we are most comfortable with a rational approach to knowledge. However, as rationality itself is based on the fundamental knowledge of direct perception, we must have at least some use of this higher faculty. As it is not possible to explain any such higher knowledge by logic, and as we are not easily able to recognise direct perception, we either reject this knowledge, or we accept it on faith. In this way faith acts as a complement to metaphor and art. We generally even accept the validity of logic and non-violence or peace in this manner without actually realising it. This is true of modern science too, which accepts the validity of logic when its own rational methods alone cannot give a proper justification of it.

The question of religion is obviously another important example of the use of faith. It is an attempt to justify our directly perceived glimpse of transcendence or reality, but in today's materially dominated culture many of us decide to reject this aspect of faith as it does not seem relevant to our existence. We are happier to accept logic and non-violence instead, as they appear to be more useful than religion in a material model of reality. However the danger of religious faith is that it lends itself to abuse when the direct perception is not well developed. Often we are asked to accept ideas which do not have the truth of direct perception behind them on faith alone. This takes us away from natural stability and unity, and so it is against the natural order or dharma. Hence it is also important to be wary of faith in something when we cannot see its purpose.

Important Vedic Sciences and their Roles

As the Vedic sciences are based on a consistent and rational foundation at a fundamental level, i.e. a foundation based on reality, each of these disciplines are capable of yielding results which can pave the way for the deeper knowledge of direct perception. They can give insights into fundamental questions which appear to have escaped modern science due to its self-referential inconsistencies.

When we consider examples of specific Vedic sciences, we should note that sciences such as Physics, Biology etc. are not automatically modern sciences. They can be practised as a modern science or a Vedic science, depending on the approach. To this extent some important Vedic sciences are:

Ayurveda – Science of Life and Health
Yoga – Science of Mental Control and Unification
Vedic Astrology (Jyotisha) – Science of Light and Time (Including Astronomy)
Tantra – Science of Energy (Shakti)
Gandharva Veda – Science of Music
Vastu – Science of Space, Land and Building Management
Vedic Mathematics
*Science of Language – Including Mantra, Grammar, Etymology, Meter, Pronunciatio*n and Sanskrit

The first four of these form the next four chapters of the book, and the Science of Language was included to a certain extent in the chapter on mantras. These are perhaps the most important of the Vedic sciences.

The Vedic sciences are eventually overlapping facets of one whole due to their basis in one reality. In fact they really form one discipline, and it is at the physical levels of manifestation that they specialise into their own areas. They are designed to address the various levels of our pentuple being as described in the chapter on Cosmology. For example, Ayurveda addresses the food and Prana bodies, and Yoga addresses the emotional and intelligence bodies. Vedic astrology brings our actions into harmony with nature's timing (i.e. it makes them dharmic actions), which works according to the law of cause and effect. Tantra similarly helps us to optimise the use of Shakti, or nature's energies and forces. The sciences of language and music help us to understand the subtle or mental rhythms of nature from a basis in causal sound (pashyanti). To this extent Vedic music is effectively a rhythmical and melodious version of mantras.

ENDNOTES

1. See Schrödinger's Meine Weltansicht, and The Agony of J. Robert Oppenheimer by Richard Rhodes. Both of them adopted the world view of Hindu philosophy. There are many quotations of Schopenhauer's extreme praise of Vedanta, some of which are in The Discovery of India by Jawaharlal Nehru. Heisenberg studied Vedanta in India at Shantiniketan. Einstein is accredited with the following quote: 'Access to the Vedas is the greatest privilege the 20th century may claim over all the other centuries'.
2. Unfortunately there is not much of his work available in English at present, but refer to Abhinavagupta by GT Deshpande for more on this.
3. For a beautiful story on the knowledge of knowledge, read the story of Shvetaketu in the Chandogya Upanishad 6.1
4. See Note 2 of Chapter 1 for some references on the limitations of logic.
5. To add to Note 2 of Chapter 1, in the early part of the 20th Century Kurt Gödel gave a modern presentation of this in the field of Mathematical Philosophy and Logic with his Incompleteness Theorem. It is technical and nature, and requires some advanced knowledge of mathematics.

6.

Ayurveda: The Science of Life

Lead me from Death to Immortality.

Brihadaranyaka Upanishad 1.3.28

Vedic Healing and Modern Medicine

Ayurveda is the Vedic science that particularly deals with the health of the three lower (physical, instinctual and emotional) koshas. It also addresses the higher koshas (intelligence and bliss), but there its techniques merge with the science of Yoga and the control of the mind. It has a deep healing power as it has a consistent and fundamental foundation in reality and rationality, an essential aspect of the Vedic sciences. This makes it particularly useful in dealing with deep seated and chronic diseases. Ultimately Ayurveda views a deviation from the reality of the unified stability of Purusha as a disease, and so it is eventually a path of Yoga. This is its deepest goal, and as such, it is more important than its specialisation in material medicine. However, it also has a highly developed system of material medicine, and it is especially noted for its use of the healing power of herbs, minerals and a healthy lifestyle.

It should be recognised that modern medicine is very valuable, with its own highly developed material approach. This is particularly so in acute illnesses where the primary cause of disturbance to the physical body is material, such as in the case of accidents, and also where surgery is required. Due to the dominance of modern medicine in this area, the skills of Ayurveda have declined in it despite being well developed in the past. However, when we go closer to the mental and causal levels of disease and look at the laws of nature there, modern medicine is not effective as it has not even recognised the existence of these levels. Ayurveda attempts to directly incorporate these levels into its healing methodology, and thereby it benefits from the intelligence of nature on a vast scale. As such Ayurveda attempts to directly observe the laws and principles of infinite nature, rather than limiting itself to human intelligence and laws. The great importance placed in conscious and intelligent reality for healing in Ayurveda is indicated in the very word for healing and therapies in Sanskrit, which is chikitsa. In Sanskrit the root of chikitsa ('chi') is the same as the root for the word for consciousness, which is Chit. So chikitsa effectively means 'with consciousness'.

Modern medicine also uses human intelligence well, but its usage of intelligence is naturally limited, because it does not recognise the existence of an independent subject or a stable reality as the basis of an otherwise completely unstable material word. Thus modern medicine, which limits itself to the material world, cannot go to the causal or real level of diseases, and so it can only heal at an outer level.

Additionally, Ayurveda is also able to reduce incorrect decision making with the appropriate usage of Yogic mental and emotional management. Incorrect decision making is a feature of the human kingdom as we saw in the chapter on Cosmology. These Yogic management tools are not available to modern medicine, once again due to the fact that it does not recognise the existence of reality apart from the material world.

As Old as Nature

As Ayurveda bases itself directly on the healing power present in nature, it is regarded as being as old as nature. This makes Ayurveda the 'mother' of all systems of healing. Even from the perspective of conventional human history we see Ayurvedic concepts explicitly being mentioned in the Rig Veda[1] itself, the oldest known book to humanity at present. At present the commonly practised version of Ayurveda is based on the Charaka Samhita (Charaka's Treatise), which is usually dated to about 2,500 years ago. To add to the age of its knowledge, its text says that it is in fact a compilation from earlier times. It is a work of a highly developed system of medicine, and it is the oldest known book of its type. To this extent, Ayurvedic influences can be seen in subsequent systems of medicine such as the Greek system (which serves as basis for modern medicine), the Arabic Unani system, the Tibetan system and the Chinese system.

The Application of the Triad as the Doshas

Perhaps the best example of the practical application of the principles of Vedic cosmology occurs in the field of Ayurveda. Before we see how that works, it is helpful to look at the way consciousness projects itself into the physical world, and in this case, particularly into the human body. We saw that consciousness by an act of its will or force (Prana) forms the basis for energy as light (Prakasha), and subsequently matter. Prana represents the flow or force field of consciousness, and these flows form the mental framework of physical manifestation. This is first done by the establishment of the five universal elements of the Sankhya system, which are really states of mind stimulated by the sense organs. The Prana force fields further establish these elements into various physical forms, including the form of the human body. Prana force fields are known as nadis in Sanskrit, and form what is known as the subtle nervous system. Points of concentration of the nadis are where important centres or organs and glands of the physical body are established. The system of the seven chakras (along with their associated organs such as the heart and brain) is an example of the concentration of the nadis.

The Ayurvedic theory of the management of the health of the three lower koshas of manifestation involves connecting the concept of the pentuple with that of the triad. The pentuple of the five elements formed by the action of the life-force of Prana in the nadis also functions as a stable triad known as the doshas. They are the lower forms of the Deva triad of Agni, Vayu and Soma, and at this level they represent the building blocks of the lower koshas. They are also responsible for building the two higher koshas (intelligence and bliss), but at their higher levels the doshas Agni, Vayu and Soma. Here they are more subtle and influence the body and mind more as pure force and energy fields, rather than building blocks. These concepts will be examined in more detail later in this chapter.

Dosha means weakness in Sanskrit, and this triad is known specifically as the doshas when they are at the level of the three lowest koshas (plus the lower rationality aspect of the intelligence kosha). It is a reflection of the fact that the lower koshas are not very stable and are prone to decay, particularly the physical kosha. The three doshas are known as Vata (relating to Vayu), Pitta (relating to Agni) and Kapha (relating to Soma). They relate to the five elements in the following manner:

Space and Air – Vata
Fire and Water – Pitta
Water and Earth – Kapha

At the level of the physical body this correspondence of the doshas is to the physical elements. At the level of the instinctual and emotional koshas, they correspond more to the tanmatras of the Sankhya system which we saw in the chapter on Cosmology. Note that the combination of fire and air is the only missing pair in this scheme. At the dense level of the body Pitta is better represented by fire and water, but at the less dense level of the mind, Pitta is better represented by fire and air. It represents the evolutionary processes of nature as Agni the evolutionary fire develops in us from lower forms to higher forms.

The Mechanics of the Doshas

As the doshas relate particularly to the physical, instinctual and emotional koshas, they play a very significant role in the practical and day to day matters of life. When also considered as Agni, Vayu and Soma, their functions span across the five koshas, and so they give a deeper connection to even the simple physical functions of life. As such, they can make daily living into a metaphor or a form of art, and they can provide very accessible and important tools in the path of evolution towards reality and stability. Ayurveda involves living in accordance with the doshas, and so it is not only the science of health, but it is also the art of health. For these reasons it is important for us to take a look at some of the practical properties of the doshas, and also to look at how we may work with them therapeutically at the level of the lower koshas. Their higher influences will be looked at after that.

The main way our physical bodies are built is by our food. As the triad is the complete and fundamental concept for building structure in the universe, the physical body also builds its structure via food in the form of a triad. We can see that the digestive system has three main components, which are the stomach, the small intestine and the large intestine. Thus the triad of the doshas enter the body via food and naturally correspond to the triad of the digestive tract from where it can further build the body. Now we can also see that the properties of Kapha dosha are heavy as it relates to the earth and water elements. It is therefore the dosha most responsible for physical structure, and it is anabolic in nature. In terms of the digestive tract it naturally connects to the stomach which is also the heaviest part of the triad of the digestive tract. Similarly Pitta is the hottest of the doshas as it is the only one that relates to the fire element, and this naturally corresponds to the small intestine, which is the hottest part of the digestive triad. The fire and heat of Pitta represents energy, and this makes it metabolic in nature. It is by the application of heat or energy that it is possible to change the state of things. Vata relates to the air and space elements, and it naturally corresponds to the gaseous part of the digestive triad, which is the large intestine. As this is where waste is rejected from the body, it is catabolic in nature. This shows us the completeness of the doshas, based on the completeness of the triad of catabolism, metabolism and anabolism.

The relationship between Agni, Vayu and Soma that we saw earlier in the book has a correspondence to a similar relationship between the three doshas of Vata, Pitta and Kapha. We can consider Kapha as the basic fuel (anabolic) for the body, Pitta as the fire or digester (metabolic), and Vata as the resulting motion of the waste substances (toxic Kapha) out of the body (catabolic).

There is also another important way to view the motion of Vata, which is as the source of Pitta and Kapha. The pure aspect of the result of the interaction of Pitta and Kapha as fuel and its activity represents an uplifting motion of Vata. The unity of the purer Vata splits into a dual as Pitta and Kapha to operate on a higher level, which is much like how Purusha appears to be a dual as a basis for manifestation via Prana, and also how Prana facilitates evolution in manifestation. Thus this uplifting or positive role of Vata relates to Prana, and it is the way the life-force spreads through the human being starting with the digestive tract to eventually nourish its deeper aspects. It is the representation of the cycle of cosmic Prana, which enters us first as food of the physical body via the manifestation of the doshas, then it is uplifted to the life-force of the instincts and emotions, until eventually the cosmic Prana of the higher koshas is once again attained.

We should note that the downward motion relates to Prana as a life giver only when it relates to Purusha and involution. The downward motion of Pitta and Kapha is more toxic in nature, and it relates to Vata as the negative form of Prana. The upwards and downwards motion of Vata relating to Prana and decay make it in charge of health and progress, as well as disease and death. For this reason it is commonly known as the 'master' dosha.

The Doshas as Pentuples

The connection of the doshas to the pentuple of five elements is also reflected in the way they relate to the physical body. Each of the doshas has five ways (sub-doshas) it controls the body, and these sub-doshas reflect the properties of the dosha concerned across the body. As such the sub-doshas are not sources of weakness like the doshas themselves, but reflect the governing properties of the doshas that enable the body to function.

In the case of Vata dosha, the sub-doshas reflect the different types of motion at a general level in the body. Firstly there is a motion that goes inwards to the body, which is known as the sub-dosha of prana. This is not to be confused with Prana as the pure life-force, but instead the sub-dosha of prana is only an aspect of the greater Prana as it brings life into the body. It is responsible for the inward movement of the breath, food and water, as well as the sensations that feed the mind. The next Vata sub-dosha is udana, which represents upward movement such as the upward motion of air and energy required for speech. At a mental level it allows the energy of food to be reflected upwards as the thoughts of the mind, and so udana is responsible for the development of our higher faculties. The third sub-dosha is apana, and it represents the downward motion of the body (the opposite of udana). It is in charge of the ejection of waste materials from the body, as well as the reproductive functions. The fourth sub-dosha is vyana, which is responsible for circulation in the body, and thus it is closely connected to the heart. The final Vata sub-dosha is samana, and this represents the motion of contraction or absorption. Therefore it is primarily active in the digestive tract, but it is also present in other areas where such a motion is required, e.g. for the absorption of the air in the lungs, or for the absorption of nutrients in the blood and in the cells.

Additionally, as the doshas have higher or more subtle functions in the higher bodies, so do the sub-doshas. For example, in the case of the emotions, the function of the Vata sub-doshas would represent the intake (prana) of emotions, their acceptance (samana), or their rejection (apana). Emotions will be looked at in a later section of this chapter.

The Pitta sub-doshas relate to the functions of heat (metabolic activity) and light in the body. The first of the Pitta sub-doshas is known as sadhaka, and this relates to the metabolic activity of the brain. The next Pitta sub-dosha is alochaka which governs the metabolism of light via the eyes. The third sub-dosha is pachaka, and this relates to the metabolism of food which is centred in the small intestine. The fourth sub-dosha is ranjaka which relates to the liver in its role as the main metabolic organ of the body. The final Pitta sub-dosha is bhrajaka which relates to the skin and its function of heat and light management for the entire body, via the sweat glands and the pigmentation.

The Kapha sub-doshas relate to the functions of the watery or oily structures in the body. The first of these sub-doshas is tarpaka which relates to the fats in the brain (e.g. Omega 3). Next is the bodhaka which is the saliva in the mouth. Third is kledaka which is the mucous lining of the stomach. Fourth is avalambaka which is the mucous lining of the lungs required for the absorption of oxygen. Finally, the fifth Kapha sub-dosha is sleshaka, which is the lubricants in the joints such as the synovial fluid.

Notice that as the large intestine is the main seat of Vata in the body, the apana sub-dosha is particularly important in the management of Vata. Similarly, the small intestine as the main seat of Pitta makes the pachaka sub-dosha important for the management of Pitta, and the stomach as the main seat of Kapha makes the kledaka sub-dosha important for the management of Kapha. We will discuss the management of the doshas in the therapies section of this chapter.

Physical and Emotional Characteristics of the Doshas

We already have the basic outline of the doshas above, but we will look at their human qualities in some more detail in order to see what they mean for us. From their connections with the five elements, we can describe their general qualities in the following manner:

Vata – Mobile, Light, Cold, Dry
Pitta – Mobile, Light, Hot, Oily
Kapha – Static, Heavy, Cold, Oily

The space and air elements that compose Vata make it naturally very light and mobile. There is no heat and water present in Vata, which also makes it cold and dry. Pitta is obviously hot due to the fire element, and it is slightly oily as the fire element requires some 'fuel' to sustain it. The combination of fire and a little water also makes it light and mobile. Kapha is heavy and static due to the earth and water elements, and cold, as it has no fire element. An abundance of the water element also makes it quite oily. The outstanding qualities for the three doshas are cold and dry for Vata, hot and wet for Pitta, and cold and wet for Kapha.

These qualities have their different effects on us. Physically, we can naturally observe their effects as the following:

Vata – Thin and light frame, prone to dryness, e.g. in the skin and hair. Dryness in the digestion can also cause constipation. Weakest of the doshas due to catabolic nature.

Pitta – Medium frame, slightly oily hair and skin. Heat and oil in the digestion can cause loose bowel movements. The middle dosha in terms of constitutional strength.

Kapha – Well built frame, oily hair and skin. Bowel movement is slow due to cold and oily nature. Strongest of the doshas due to anabolic nature.

Similarly we can see the effects of the doshas on the emotional body as below:

> Vata – Volatile and active in nature, prone to fear and anxiety.
> Pitta – Steady and active in nature, but angry and competitive.
> Kapha – Inactive in nature, accumulative and greedy.

These qualities are directly based on the more subtle qualities of the elements, or the tanmatras which were discussed in philosophy of the Sankhya system. For example, in the case of Vata the motion of air is active and volatile. This naturally leads to instability and fear. For Pitta, heat is active but not volatile. Heat is also sharp and penetrating like anger. For Kapha, the dense elements lead to an inactive nature. These elements naturally relate to a sense of dense material accumulation, or greediness.

The above emotional traits are negative in nature generally as the doshas are factors of weakness and disease in the lower koshas. We shall see their more positive traits when we consider them from the perspective of their higher forms later in the chapter.

Karma and the Concepts of Health, Disease and Reincarnation

Before it is possible to work further with the information we have acquired from an examination of the doshas, we need to see how they relate to the factors that cause health and disease. Then we can show they can be used for our health, progress and evolution.

First of all, we observe in nature that nothing ever occurs twice in space or time. We can also reach this conclusion from the point of view of rational principles. The law of cause and effect requires that the effects are different to the causes for it to be able to work, or else there would just be a cause and no effects. This means all subsequent effects of a cause, however far down the chain of cause and effect in time or space these effects maybe from the original cause, must be different to that cause. This is because all the subsequent effects of are cause are equivalent to one effect of the original cause. From this we have that a state of manifestation (or anything) can never be repeated regardless of how far it may be removed in time and space from the cause. If it was repeated, it would imply a break down of the greatest law of the manifest universe, which would also mean the break down of the manifest universe. Time itself would not exist, as we mark time by changes, and changes occur via the law of cause and effect. Thus it is possible to rationally derive the truth of what is also the easily observable fact that everything is unique, whether it is a grain of sand, a rain drop, or a human being.

Notice that if we consider manifestation as based on Purusha, and also evolving to Purusha, we have cause and effect as the same thing, i.e. Purusha. If cause and effect are the same, then nothing is happening which means a break down in manifestation. It is equivalent to saying everything is constantly repeated as one thing. This is indeed true from a fundamental perspective, and it is another way of presenting the philosophy of advaita which says in reality (or in Purusha consciousness) manifestation is not real, but only appears to be so as the phenomenon of Maya. However from the perspective of manifestation which is not fully based on reality, i.e. a human perspective, we see the law of cause and effect working conventionally, and we have to accept manifestation and Maya along with their consequences.

These consequences include the concepts of health and disease. Manifestation occurs in time, and anything that exists in time must also end in time. This involves its decay and death, or disease. Ultimately health and evolution is realising we really exist independently of time (Purusha), where there can be no concept of death and disease. Thus in

Ayurveda it is said that disease is a failure of intelligence, or due to not realising Purusha consciousness.

The concept of uniqueness in manifestation leads to the conclusion that the triad of the doshas must make us unique as humans. The triad allows for an infinite variety in manifestation generally, which is achieved by the variations in the ratio of each of the three doshas in the triad. Therefore the doshas have a unique ratio in each of us. The ratio we are born with is the natural ratio of our lower koshas, and so it is our position of equilibrium or health at the level of the lower koshas in our lives. From the perspective of the analogy of Agni, Vayu and Soma and the upward and downward triangles, it is the triangle our lower koshas are at in our lives before they can evolve up to a higher one. Maintaining this ratio is termed as being healthy and facilitates evolution to a higher triangle, and a departure from it results in disease and descent to a lower triangle.

The unique ratio of the doshas for each of us is determined by our prior actions at the level of our lower koshas (up to the level of rationality) in accordance with the law of karma, which is the law of cause and effect generalised to all the koshas. These koshas are thus maintained, destroyed or renewed according to the consequences of the actions at their level. The law of karma is highly complex when compared to the simpler physical law of cause and effect, because in the non-physical koshas the law of cause and effect is not necessarily of the simple linear form we see in the physical world. The less dense structure of these koshas allows for a complex intertwining of manifestation at these levels (as we noted in the quote from the Brihadaranyaka Upanishad earlier), and a similarly complex unfoldment of karma. It is what brings the infinite complexity we see in the world, and it is also the rationale behind the concept of reincarnation, as described below.

The law of karma requires that effects of causes that are yet to unfold are not lost, but that they are stored in a latent form, much like how a seed stores its potential. These seeds are called samskaras in Sanskrit, and they determine our future attributes, strengths, weaknesses etc. They form the undercurrent of our future actions as desires (vasanas), and these desires then manifest as our activity or effects at the outer level. The samskaras are stored at the level of jivatman (relating to the soul from the chapter on Cosmology), as the jivatman connects to our eternal aspect which lies beyond time and karma. Thus it is at the level of the jivatman that the evolution of the koshas is determined, and it is the basis of the theory of reincarnation. It is a direct result of the inviolable and fundamental principle of cause and effect, which also determines our attributes at birth. These must be unique as the same human personality can never reincarnate again (as nothing can ever be repeated), but only one that is karmically related to it. When such karmic results unfold during a single life time, we generally refer to them as being due to 'lifestyle'.

Diagnostics

Diagnostics in Ayurveda is concerned with determining our natural dosha ratio, as well as the nature and extent of any deviations away from it. There is a vast array of techniques for this purpose in Ayurveda, and they are based upon detecting the natural qualities of the doshas. Normally these techniques involve determining our general characteristics as individuals, plus examining our digestive systems, the channels of flow or movement (which relates to Prana) in our lower koshas, and the level of penetration and manifestation of any disease condition that maybe present in us. Diagnostics is a more technical area, and as such, it is more suitable for those who are interested in a further study of the discipline.

Once a diagnosis has been established, maintaining or regaining our natural dosha ratio is the field of therapies and remedies in Ayurveda. As therapies affect our way of living the next sections look at them in more detail.

Ayurvedic Therapies

Ayurvedic therapies are usually divided into three main approaches. The first approach is known as the intellectually or rationally dominated approach, because it deals with the three lower koshas which are usually controlled by the rational intellect. As the lower koshas are the main area of activity of the doshas too, the intellectual approach is the primary approach in managing the doshas. Its therapies mainly concentrate on the physical body, and it affects the other lower koshas via their connection to the body.

The next approach deals with the higher koshas, and so it relates more to the Devas and is known as Daivic or divine healing. It attempts to add the healing power of direct perception to that of the intellect. However can it include the various other non-physical therapies across the koshas too. Finally the third approach is one that directly aims for Sattva, the universal quality of balance, and gives the doshas a lower priority. These last two approaches will be looked at in later sections of the chapter. The intellectual (or lower therapies) approach is the most common approach of Ayurveda as it is mainly concerned with the lower koshas and the doshas. The other approaches are where Ayurveda and the methods of Yoga come together, as the main focus of Yoga is the higher koshas.

Lower Therapies

We saw the key to Ayurvedic therapies lies in maintaining the natural dosha ratio of a person, and where this ratio is imbalanced, restoring the natural balance. In the case of the lower therapies this process is managed intellectually and rationally in Ayurveda, by using the principle that like increases like, and opposites decrease opposites. We shall see this principle outlined relative to the dosha qualities in some of the therapies described below. Importance is given to control any dominant doshas in the natural balance, as it can easily dominate the other doshas and disturb the natural balance. It is thus usually the main source of disease in a person, along with Vata dosha, which we saw rules over disease and decay in any case. The most important tools for the implementation of the lower therapies are diet, herbs, minerals, exercise and detoxification.

The maintenance of the natural dosha balance usually comes under the approach of lifestyle management in Ayurveda, whereas the rebalancing of an imbalance in the natural dosha balance more commonly comes under the approach of disease therapies. Both of these methods involve the concept of creating 'clean fuels' for us according to our natural dosha balance. The qualities of the doshas have been given in some detail in Ayurvedic texts, and so there is a great deal of guidance and advice available as far as lifestyle goes. In Ayurveda an appropriate lifestyle was emphasised from the most ancient of times, in contrast to modern medicine which has to begun to acknowledge the importance of lifestyle measures only relatively recently. It is an indication of the power of the Vedic approach of direct perception to knowledge. Lifestyle management can have a lower and a higher approach, and in the next section we shall consider the lower lifestyle methods. In the section on Higher Therapies we shall consider the higher lifestyle methods.

Another way of analysing the lower therapies is by considering them as building or reducing. The building approach is given when a person is physically weakened and needs rejuvenation and tonification, and is concerned with doshas that have become weaker than required for the natural balance. It is more common for Vata types as it is the weakest dosha. Building is also used after strong reduction therapies, which can a leave a person a little weak for a while, as we shall see later. The reduction approach involves reducing the doshas that exceed the natural balance. Reduction itself is of two types, strong and weak. Weak reduction is also known as palliation, and is practised as a part of the lifestyle approach (as can some aspects of building). Strong reduction is usually a clinical procedure lasting a few weeks, and consists of five main ways (panchakarma) of detoxifying the body from any excess doshas. Hence this relates more to the disease therapies.

Lower Therapies - Lifestyle

The lifestyle approach uses the principle that opposites reduce opposites, whether for palliation or tonification. Typically palliation is used for Kapha and Pitta doshas, and tonification for Vata dosha as it is a naturally reducing dosha, but this depends on the specific circumstances. For chronic conditions that last a long time, even Pitta and Kapha types can become weakened, and may require a combination of palliation and tonification. Conversely, it is possible for Vata types not to require much tonification for shorter term conditions. The lifestyle procedures of palliation and tonification are generally mild, and are equivalent to maintaining the natural dosha balance, i.e. they are generally used where there is no significant imbalance or disease already. However, it is possible for them to have a similar effect to the stronger procedures if they are practised over longer periods, such as a year or more.

When using the opposites principle for the lifestyle approach, it is necessary to look at the dosha qualities as described above, and then implement a lifestyle that compensates for these qualities. For example, when it comes to looking at the diet, Vata would benefit from heavier and warmer foods, Pitta from cooler foods, and Kapha from lighter and warmer foods. On the basis of this we could recommend the following diets[2] for the three dosha types:

> Vata – most grains, dairy products, cooked vegetables, heavier fruits, oils such as sesame oil and ghee, seeds and nuts, moderate use of herbs and spices.
>
> Pitta – most grains, dairy products such as milk and cream, sweet fruits, cooked and raw vegetables,

limited use of oils except ghee, mild use of herbs and spices.

Kapha – cooked and raw vegetables, light fruits, limited grains such as whole grain rice, barley and mung beans, heavier use of spices.

Vata types also benefit from a regular (e.g. 5 times a day) eating frequency (though in smaller quantities due to their lighter structure), Pitta types from a moderate frequency (3 times a day), and Kapha types from eating infrequently (2 times a day).

The above is just meant as a limited introduction for illustrative purposes, and there are many good Ayurvedic books available that give much more detailed information. Some of these have been recommended in the Bibliography and Notes sections at the end of the book.

Similarly, we can also outline some activity and behaviour patterns suitable for the doshas as a basis of the non-physical aspects of a lifestyle:

Vata – relaxed pace of activity and exercise, e.g. conventional Yoga asanas. Emotionally, peace, faith and courage should be cultivated.

Pitta – relaxed peace of activity and cooling exercises such the gentler Yoga asanas, and walks along water bodies or in the hills. Emotionally, forgiveness and compassion should be cultivated.

Kapha – stimulating pace of activity and exercise, such as the more vigorous Yoga asanas, trekking etc. Emotionally, contentment and detachment should be cultivated.

We can see that peace and faith can overcome the volatility of Vata, compassion can overcome the aggression of Pitta, and detachment can overcome the greed of Kapha.

We should note that the lifestyle approach above is particularly suitable when a person is close to their natural dosha balance, but it should also take into account other factors such as the seasons. Where there are strong seasonal effects, these should be given similar or greater importance than the natural dosha strengths and weaknesses for a person in relative dosha balance. For example, in places where there are extremely hot summers, most people would benefit from a lifestyle that is mainly anti-Pitta during that season. Similarly the converse is true for places with very cold winters.

Lower Therapies - Strong Reduction (Panchakarma)

The strong reduction therapies consist of a three staged process based on herbal medications. The first stage is a preparatory stage, and the second stage is the actual reduction or detoxification stage which normally has five main ways (panchakarma) of being implemented. The third stage is a follow up stage to rejuvenate the body after the strong reduction. The panchakarma procedures are used when the natural dosha ratio becomes imbalanced and there are disease symptoms being manifest. In this case the lifestyle methods above may not be sufficient, or may take too long to restore health. The three staged approach is of great importance in the success of this type of treatment as we shall see, and it is based on the concept of the upward pointing triangle (a triangle based on clean fuels).

The doshas have their main sites in the digestive tract, and hence it is natural to implement the detoxification at these sites. However if we do this without the preparatory phase, it just simply cleans up the digestive tract and not the other parts of the physical body and koshas that feed of it. When the natural dosha balance is disturbed, the doshas spread from the digestive tract to other parts of the body as toxins, and disease begins to be manifest. This requires a cleaning of the entire body, and therefore the preparatory stage involves heating and oiling (clean fuel and clean fire, or Agni and Soma) the body both internally and externally. This has the effect of increasing the metabolic rate of the body, which helps to burn or weaken any toxins that may be present in it. The oleation also increases the lubrication of the body, and this enables toxins that may have spread from the digestive tract and become lodged around the body (due to the increase in the doshas), to return to the digestive tract reasonably easily. As a result, any detoxification of the digestive tract that occurs after these preparatory procedures represents a thorough detoxification of the body, or a strong dosha reduction (clean motion or Vayu).

The panchakarma procedures involve reducing the doshas or expelling them from the digestive tract. For removing Kapha from the stomach, the closest exit from the digestive tract is the mouth, and thus reducing Kapha consists of therapeutic vomiting. There is also the nasal route which can be used to remove Kapha as mucous from the sinus cavities, and this is known as nasyam. For removing Vata from the large intestine, the nearest exit is the anus and thus this procedure involves enemas. For removing Pitta from the small intestine, which is in the middle of the digestive tract, it is best to the use natural exit or the anus as opposed to the mouth. Hence the procedure here involves laxation. The final or fifth panchakarma procedure is another way of releasing Pitta, which involves blood letting. The blood is the main way of distributing energy from the digestive tract around the body, and therefore releasing blood reduces Pitta.

As these panchakarma procedures are strong, it is essential they are done with proper care and diagnosis, and under clinical supervision. For this purpose there are many Ayurvedic hospitals around India, as well as some in other parts of the world.

The third and final stage of the strong reduction therapies is not only to rejuvenate after the reduction, but also to help maintain a balance for the higher level of purity and dosha balance of the body. The three stages form a triad as an upward pointing triangle. The preparation is like the Agni stage (metabolic), the detoxification is like the Vayu stage (catabolic), and the rejuvenation is like the Soma stage (anabolic).

The Theory of the Six Tastes

Ayurveda uses its theory of taste[3] in order to classify foods, herbs and medicines according to their effect on the doshas. The concept of taste (rasa in Sanskrit) is a very subtle one in Vedic thought. Apart from the taste of the tongue, it includes concepts such as the 'taste' in the three aspects of the digestive tract (stomach, small and large intestines), and especially the 'taste' in the mind. The mental concept of taste is like the mood, emotion or the subtle essence of food (or any substance generally). In this way even food can become like music, mantra or art, and is connected to direct perception. The various aspects of rasa allow for a very deep classification of edible substances, and as a result, Ayurveda has an extremely comprehensive and sophisticated literature in this regard from ancient times itself.[4] Much of this information is only now being discovered and verified by the methods of modern science.

The Ayurvedic theory of taste recognises six tastes, which correspond to the five elements. These six tastes are sweet, salty, sour, pungent, bitter and astringent. From their correspondence to the elements it is possible to determine which tastes are suitable for each of the doshas. The tastes of sweet, salty and sour are considered to correspond to the heavier and warmer elements and thus are suitable for Vata. Sweet, bitter and astringent are considered not to contain the fire element, and are thus suitable for Pitta. Similarly, pungent, bitter and astringent are considered to be lighter and more stimulating, and are thus suitable for Kapha. We can observe these correspondences between the doshas and the tastes in the diet table outlined earlier.

Higher Functions of the Doshas and Alchemy

The doshas' main field of activity is the three lower koshas, and they are the causes of diseases there. However, as we move to the next two koshas of intelligence and bliss, their properties change. They are no longer sources of weakness, but are positive in function and give strength. They become

Devas in us at this level, and connect us to the universal rather than the individual, in their forms as Agni, Vayu and Soma. As a result the domain of the higher koshas is no longer one of disease, and we can trace a path through the five koshas to see how the doshas evolve from their lower functions to their higher functions.

To this extent we can notice that the koshas are not discrete categories, but are in fact highly interconnected. One such connection is that a kosha forms the basis for the next kosha above it, provided the lower kosha is functioning harmoniously (based on clean fuels). For example, the physical body is the field upon which the instincts act and get their meaning. The instincts provide the sensations which form the basis of our emotions such as desire or competitiveness. The emotions in turn form the basis of our personality and identity, and this enables us to question our own nature. This then leads to the field of the intelligence kosha, and with proper discrimination and use of the intelligence, we realise who we are. Finally this leads us to the bliss kosha which is connected to reality or Purusha. When the fuels are not clean, and the functions of the koshas are not harmonious, then there can be downward movement as the koshas form the basis of the kosha below them.

The fundamental purpose of Ayurveda via its many therapeutic approaches is to facilitate this evolutionary path through the koshas. It is to ensure that clean fuels are being made available to us so that we may evolve upwards, starting with the physical body. Then we have the evolution of the doshas from their lower functions to their higher ones. This change specifically takes place from the level of the intelligence kosha. Here there is a move upwards from logic and emotions to functions that relate to direct perception. Each of the doshas affects this process differently. For instance, Vata, which is noted for volatility in the lower koshas becomes more positive by enabling a rush or flow of ideas. This underlies the inspiration of the intelligence kosha, and in the Rig Veda it relates to the metaphors of Devi

Sarasvati,[5] the river which flows with a mighty current. Similarly, the connection Pitta has with light and the ability to see or work thing out, gives us deep insight or visionary and revelatory experiences. This relates to the metaphors of Devi Ila of the Rig Veda. Kapha relates to the denser elements associated with the senses of smell and taste. At a higher level this is the ability to be intuitively drawn towards, or to desire (the main Kapha emotion), the truth. It is like the forces of magnetism and gravity which naturally attract or desire. It relates to the metaphors of the Devi Sarama of the Rig Veda.

The junction of the doshas between their higher and lower functions occurs at the level of rationality and the emotions. Vata represents a flexible mind, Pitta represents the steady activity of the mind as logical thought, and Kapha is the nurturing or caring aspect of the mind. The evolution of the doshas from the lower functions to their higher functions is the concept that underlies Vedic alchemy. Ayurveda shows us how even the physical act of eating food can be transmuted to an action which can give us divine or Deva powers such as inspiration, revelation and intuition. Conversely, we must be aware that poor physical acts can lead to demonic behaviour at a relatively higher level.

Higher Therapies

The higher therapies of Ayurveda ultimately focus on the concept of the healing power of the Devas and Devis. In other words, they attempt to harness the healing power of direct perception in addition to the intellectual approach. In their widest definition they can refer to all the non-physical therapies of Ayurveda across all the koshas, though their most powerful methods such as mantra and meditation do focus on the higher koshas. As we shall see, several of these therapies are essentially a 'higher' lifestyle approach.

Starting with the instinct kosha, this approach consists of providing the instincts with their natural stimulation or food bearing in mind the prevailing dosha conditions. The natural stimulation by dosha is as follows:

Vata – Calm and stable environments such as lakes, valleys etc.
Pitta – Peaceful and beautiful environments such as lakes, mountain views and flowers.
Kapha – light and stimulating environments such as mountain tops and fast flowing mountain streams.

As regards the rational intellect (emotions were already addressed in the lifestyle approach), the therapeutic approach by dosha is as follows:

Vata – developing the ability of a versatile and flexible mind.
Pitta – developing the ability of rational thought.
Kapha – developing mental stamina and persistence.

This develops the positive side of each dosha, and acts as a bridge to their higher functions.

Next there are therapies relating to natural light, gems and aromas. They serve to feed the life-force in the brain and mind via their connection to the senses such as smell and sight. As smell relates to the earth element, the tanmatra relating to aroma has a particularly foundational effect on the mind. We will see some more of light and gem therapies in the chapter on Vedic Astrology. There is also the Pranic healing approach relating to the life-force for any of the koshas. At the level of the instincts this is mainly done via the breath (which is commonly known as pranayama) and the marma points.[6] Marmas are special points around the body where the subtle nervous system (nadis) anchors into the physical body. Their proper maintenance via massage (including subtle massage) is of great importance for the healthy function of the physical life-force.

At the level of the higher koshas Pranic healing can be considered to be a part of meditation and mantra practices, and these are the highest and most powerful of the Ayurvedic therapies when used to their proper potential. The energies behind these therapies are Devis such as Sarasvati, Ila and Sarama, and they uplift the doshas to their higher functions (based on the proper preparation of the doshas in the lower koshas). Here the decay of Vata is replaced by the uplifting and evolutionary Prana. The concept of balance between these higher functions is not as vital as in the case of the doshas, because they are sources of strength and an excess of any of them is not normally damaging. The meditation and mantra practices take Ayurveda into Yoga and Vedic astrology, and they will be discussed more there.

There are also higher practices that relate more directly to energy or Shakti. They work with mantras too, and also with the concept of kundalini and the chakras. This is a more tantric approach and will be looked at in its own chapter. Generally, the universality of the therapeutic approach of Ayurveda is expressed in its maxim that 'anything in the universe can be used for healing, provided we know how to use it appropriately'.

The Three Gunas and Yogic Ayurveda

The final healing approach of the three introduced at the beginning of the therapies section is the one that relates to Sattva (balance). We saw that the concept of the triad also naturally relates to the three main qualities (gunas) in manifestation, which are Sattva, Rajas (activity) and Tamas (inactivity). Various combinations of these three qualities are required to form the framework for the doshas to act as the triad of the building blocks of manifestation.

In Sattvic healing priority is given to developing Sattva over the doshas. Therapies relating to the lower koshas are dominated by the physical body, which is the densest of the koshas. The body is thus relatively the most inactive and structured, or the most Tamasic of the koshas. The doshas naturally work best here as they are the basic building blocks of manifestation. When we consider the emotional kosha,

Rajas starts getting more prominent. At the level of the higher koshas (the domain of Yoga), we see more Sattvic qualities. Higher functions such as inspiration, revelation and intuition require Sattvic balance, as we shall see in the chapter on Yoga.

From this we can see that the doshas have their full variety of manifestation from the lower koshas upwards, whereas the gunas get this only at the level of the higher koshas. Therefore for the higher koshas it is necessary to consider the gunas in addition to the doshas, and as a result the gunas form the basis of psychological healing in Ayurveda.

The Sattvic approach attempts to develop balance from the level of the lower koshas, with the aim of developing it in the higher koshas. However, in most cases this will not be very effective without using the preceding and accompanying dosha approach. The health of the higher koshas depends on the health of the lower koshas for a foundation, and the lower koshas themselves primarily depend on the doshas. To this extent, Yogic practices traditionally took into account the doshas first. Developing the Sattva of the higher koshas is also the goal of Ayurveda in any case, and this is the concept of Yogic Ayurveda.

ENDNOTES

1. For example Rig Veda 10.186 is a hymn to Vayu and Vata by Rishi Vatayana.
2. For much more detail, refer to the many good Ayurvedic books available in English now, e.g. Ayurvedic Healing by David Frawley and The Ayurveda Encyclopaedia by Swami Sada Shiva Tirtha
3. For an explanation of this see Yoga of Herbs by David Frawley and Vasant Lad.
4. For example refer to the Bhavaprakasham by Bhavamishra.
5. See Secret of the Veda by Sri Aurobindo.
6. See Marma Therapy, by David Frawley, Subhash Ranade and Avinash Lele

7.

Jyotisha: The Science of Light (Astronomy and Astrology)

Lead me from Darkness to Light.

Brihadaranyaka Upanishad 1.3.28

Meaning and Purpose

Jyotisha is the science of light, and it is known as astronomy where it concerns physical light, and it is known as (Vedic) astrology where it concerns the light of the mental levels. Perhaps the best example of the emphasis on the eternal and the permanent in the Vedic sciences is in Jyotisha. It begins with the permanent 'I am' awareness we have at the inner level, and brings it together with the most permanent awareness we have at the outer physical level, which is the light of the stars. It uses its many techniques which are based on the light of the stars to bring clarity to the implications of the light of our inner permanence, at the outer level of our daily lives. Some of these techniques will be illustrated in this chapter.

Jyotisha is ultimately the science of the stellar light of consciousness (Prakasha). It is this light that gives us the insights of direct perception, and as such it is considered the 'eye' of the Vedas. For this reason it is also considered as the foremost 'limb' (Vedanga) of the Vedas. The Vedangas are auxiliary disciplines to the Vedas whose specific purpose to bring forward the deeper meaning of their verses, or to develop direct perception. As direct perception is especially aided by the metaphors of art, poetry and other such inspirational aids, the other Vedangas are concerned with the nature of language and poetry. In this manner they help us to directly see the verses of the Vedas by the light of Jyotisha.

This is also why the Vedas or any of the associated disciplines such as music, Ayurveda, Yoga etc., were originally maintained as oral traditions. An oral transmission of knowledge helps facilitate direct perception via the living inspiration of the teacher to the student, as the oral word is closer to consciousness than the written word. With the decline in the ability to directly perceive (for reasons we will see later in this chapter in the section on Yugas), the Vedas were presented in written format. It helped preserve their structure where the level of direct perception on its own was not sufficient to do so. Just as the level of our insight into the Vedas has been reduced, so has the depth of our usage of Jyotisha. However, much as we can use audible mantras to connect us to causal levels of sound and consciousness, we can use a more logical and physical form of Jyotisha even with its reduced level of direct perception, to connect us to the highest levels of light and perception. This is the main scope of the practice of Vedic astrology today, and as a result astronomy also forms an important part of it.

Much like the highest levels of Jyotisha give us access to the deepest laws inherent in consciousness and the universe, the outer forms of Jyotisha give us access to the laws that govern us as individuals. In the physical world we see light as defining time as will be explained in the next section. All the laws of the universe operate in time, and hence a good understanding of the measurement of light enables to see how these laws, especially how the law of cause and effect, relate to us. Jyotisha does give us a good understanding of this at an individual level as we shall see, and thereby it allows us to harmonise our actions efficiently with nature's laws, which is the concept of dharma. Depending upon the level of direct perception we are able to use, we are able to extend this understanding to the emotional and the higher levels of our being. In any case this type of understanding at any level is naturally very useful for the purpose of our individual evolution, and it helps to give us a dharmic framework for our lives.

A Rationale

As Vedic astrology is a Vedic science, it must have a basis in rationality, with direct perception enabling to go to a deeper level. It is important to look at this rationale, as astrology is a subject that is often misunderstood. This is partly due to an over-usage and a poor practice of the discipline, and partly due to its subtle nature. A rationale also better helps us to understand and implement this important discipline for evolutionary development. For these purposes we shall outline a rationale for astrology, based on the principles as outlined below:

1. Light as a basis for time
2. Space/Time moments acting as seeds for all that is possible in that moment
3. Cyclical nature of time
4. The law of cause and effect, which works in time
5. The precedence of consciousness and the mind over physical manifestation
6. The relationship between a mental subject and physical manifestation as an object

The first principle has been recognised by modern science via Einstein's theory of relativity. Likewise, the second principle has been identified in the theories of Quantum Mechanics. The third principle has also been established via Einstein's theories, where light is shown not to travel in straight lines. We can also directly observe it in nature, where everything appears to follow cycles, whether it is day and night, the seasons, life and death etc. It is also being reflected in modern cosmology with the 'Big Bang' theory which holds that the entire universe, including time, appears and disappears in cycles. The fourth principle is obvious via direct perception generally, and the validity of the fifth and sixth principles has been discussed extensively in earlier chapters of the book.

First we shall see how these principles, which relate closely to modern thought, together allow for a rationale for astrology. After that we will also establish a rationale directly from the perspective of Vedic cosmology.

First Principle - Relativity theory of modern Physics tells us that light defines time. The one feature that all physical objects of manifestation have in common is that they either are a source of light, or they reflect light. Hence we can primarily define objects of manifestation by their light (note that Black Holes are the obvious exception to this, but they remain indefinable in physical existence). The changes in physical manifestation represent the passage of time, and these depend on the changes of light emitted by the physical objects of manifestation.

The main source of light for the Earth is the Sun, followed by the reflected Sun light of the Moon. From very ancient times to the present day they have remained the two most important objects for the measurement of time, defining the day, month and year. The next most important sources of light on the Earth are the planets (reflecting the Sun light like the Moon), plus the stars whose light is able to reaches us. These sources of light help fine tune the measurement of time, particularly when we consider longer time cycles such as the (approximate) 24,000 year cycle of the wobble of the Earth's axis of rotation. They allow for the differentiation between years, as well as millennia in these longer time cycles. This differentiation becomes particularly relevant for old and long-lasting civilisations such as the Vedic civilisation, which has time-spans of 10,000 years and more. Hence there is no better way of measuring time than observing the heavenly bodies.

Second Principle - The most meaningful way to capture the state of a particular moment in time would then naturally be to catch a snap-shot of the position of the heavenly bodies at that particular moment. There is another important factor that reinforces this point. The longer an object lasts for, the better it is able to reflect the status of a system at any moment. For example, a stone that has been present in a particular environment for several years will reveal more about that environment at any moment than a stone that has been present in that environment for just a few days. The heavenly bodies are the most ancient objects we can observe. Therefore this also makes them highly appropriate to record the state of any moment in time.

Quantum Theory requires that each moment in time must contain in latent form all the possible outcomes in that moment. From this we can see that any moment contains all the information on the entire life beginning at that moment. Moments occurring during a particular life will contain less information, as they will be restricted to the circumstances prevailing during that moment of that particular life. This is the field of Prashna or Horary Astrology. Thus the horoscope of a moment becomes the best piece information to indicate to us what is possible for a life born during that moment.

Third and Fourth Principles - The periodicity of the motion of the heavenly bodies gives us the periodicity evident in Astrology and time, relating to the third principle. This is an important factor that helps facilitate the predictive nature of astrology. The changes that occur in physical manifestation that define the passage of time obey the law of cause and effect (fourth principle). This gives the ability to deduce at least some outcomes that are possible in each moment when used in combination with the second principle as the basic starting position. It should be realised that the law of cause and effect, like principles five and six, represents an inner principle or an eternal truth, rather than being dependent on finite sensory data.

Fifth and Sixth Principle - Once we give precedence to mind over other manifestation, we can see that physical objects are in fact states of mind, or created by the mind (dream-objects). It seems reasonable that the most long-lasting and massive physical objects (i.e. the heavenly bodies) will reflect the mind (as the subject) most fundamentally as its archetypes. An example of this is the role the Sun plays in astrology. As the primary source of light and energy for the Earth, and thus the main driver of physical change and metabolic activity, it represents similar qualities for us mentally on an individual level. On an outer mental level it relates to our outer will and personality, on an inner level it relates more to our inner will and our higher koshas as the Soul. This gives us insight as to how the heavenly bodies reflect us individually at all levels of our being, and it relates to principles five and six. As the law of cause and effect governs heavenly bodies too, it shows how this law also works at the mental levels, and how it gives the ability to make predictions across the koshas of the mind and body.

We can also apply a rational approach to the Vedic theory of cosmology and arrive at a rationale for astrology from its fundamental perspective, in addition to the philosophical insights we have already seen from it. We saw in the chapter on Vedic Cosmology that the manifest universe is built not only from force, light and energy as is also recognised by modern science, but also from time and intelligence. These factors are infused through out the particles of even the physical matter in the universe. Hence the physical universe has the secrets of all aspects of time embedded in it. Modern science does recognise the connection between light and time, and light (as energy) and matter, but most modern scientists do not take this to its conclusion. We saw that modern science extrapolates physical states into the past and future from present states, but this is restricted to a material level. Vedic cosmology indicates it is possible to go further, and to extrapolate mental and more fundamental states from physical states. As the stars and planets are the largest (and ultimately only) source of all the matter, energy and light of the physical universe, it once again gives us a rationale for astrology.

Natal Astrology and Dharma and Karma

Putting all of the above points together, the best way to capture the information of the entire life of a person is via the star and planet positions at the time of their birth, or the natal horoscope. Via the workings of the law of cause and effect through time, the picture of this seed moment allows us to deduce and see future effects at the physical and mental levels of the life of a person. The circular or periodic nature of time further aids this.

As discussed earlier, the law of cause and effect can be seen to have a deeper and wider meaning to cover fundamental states and the mind, in addition to the modern understanding of the law in the physical sense. On the other hand it is a part of the physical structure of the cosmos, and not just an abstract law. This brings it in line with the meaning of the Vedic law of cause and effect, or the law of karma. The horoscope can thus be seen to be a cause and effect map in the sense of karma. Living a life that facilitates smooth execution of the law of karma is the individual dharma, or an intelligent, healthy and 'lawful' way of living that aids evolution. This is the most important purpose of Vedic astrology. Living in ignorance of, and especially contrary to this law, causes our karma to forcibly change our lives which usually requires pain and suffering.

An important point to realise is that there is no such thing as an automatically bad or good horoscope. A dharmic life allows any horoscope to give good results, and vice-versa. The Yogic dharma specifically gives us the practical tools to see goodness in all, even the so-called 'bad', and therefore it constitutes the various remedies or therapies of Vedic astrology. We saw this was also the way of Ayurvedic alchemy. Vedic alchemy is generally the way of all Vedic thought, especially Tantra, as we shall see.

Freewill and Predestination

When we consider karma and dharma from the astrological perspective, it is also relevant for us to consider the concepts of freewill and predestination. For instance, if there is freewill how can we be bound by our karma? On the other hand, if there is predestination how can we be free to live dharmically? In other words, surely karma implies the role of dharma is limited, and dharma implies we are not bound by our karma. It seems natural that only one of these two seemingly opposed concepts of freewill and predestination must be true, and this also does appear to undermine the positions of karma and dharma.

We have seen that reality is a state of perfect balance, and so all such contradictions must be ultimately resolvable. The Vedic perspective on this indicates how this might be, and it says that the contradiction and its resolution depends on the state of mind of the observer. For example if the observer feels that there is only a material world, then everything is predetermined by the laws of materials science for them, and the role of dharma would indeed be limited. However we have seen that the observer must be subjective and apart from the material world, and a material only model is inherently unsound. As a result we naturally recognise ourselves as individual subjects distinct from the objective world, and we feel free to act due to our subjectivity. Thus we naturally recognise dharma, but we do not feel our subjectivity to be bound by karma. We generally find this confusing as the material scientific model of modern science does not allow for freewill, but only predestination or karma.

Taking the Vedic perspective further, we have seen that we are ultimately unified as one, first as Ishvara in time, and then as Purusha beyond time. At the level of Ishvara the concept of freewill and predestination must merge, as there is only one thing. Here we consider that every thing is determined by the freewill of Ishvara, and karma and dharma become equivalent. In fact karma is the first dharma, and vice-versa. When we are in the state of reality beyond time, the concepts of freewill and predestination, or karma and dharma, obviously cannot exist. This is the ultimate position as ultimately there is only reality.

The implications of this are that due to our feeling of individual subjectivity at the level of the emotions and the intellect, we must indeed exercise our freewill as dharmic activity. The dense material world, which includes our physical bodies gives the appearance of being predetermined and governed by just the material version of the law of cause and effect, until we recognise the role of consciousness and subjectivity in its existence. The more material or inanimate a thing, the more it appears to be bound by this law. At the causal level our individual subjectivity of the level of the higher koshas is determined by the will of (and is synonymous to) Ishvara as karma. Hence at the level of the formation of our individual subjectivity at the level of the bliss kosha (i.e. across the span of physical lives), we are bound by karma.

The existence of such paradoxes in nature is once again a reflection of the self-referential nature of rationality. To this extent rationality is eventually limited to telling us in this case that freewill exists because it is predestined, or vice-versa, that predestination exists because it is freely willed. We can provide a rational framework for the resolution of these type of paradoxes as indicated in this section, but ultimately a complete resolution for them requires direct perception as we saw in the case of Brahman and Maya in the chapter on Vedanta.

The Origins of Astrology and Astronomy

As with most Vedic knowledge, it is possible to find the origins of Astrology in poetic form in the Rig Veda, the oldest text we have today. The basis of the zodiac is set out there,[1] and the connection between star light and knowledge is also evident in it. It is here that Jyotisha (stellar light) is described as being the very eye of the Veda (knowledge). The 12 fold division of the zodiac that we have today relates to the 12 Rig Vedic Adityas described in the chapter on Devas. This group of Devas specifically relate to stellar light and the 12 months of the year. There are also several other references to 12 fold divisions of wheels and circles. Additionally, importance is given to the beginning (zero degrees) of the constellation of Aries, which is traditionally used to mark the beginning of the zodiac.

Apart from the 12 fold zodiac which relates more to the Sun, references to a separate definition of the zodiac for the Moon also occurs in the Rig Veda.[2] We see names of constellations (nakṣhatras) of such a 27 fold division of the zodiac occurring there for the first time. The period the Moon takes to complete one full revolution of the zodiac is 27 days, and this is perhaps the most prominent feature we can observe in the sky in relation to the stars as the Moon is the brightest object in the night sky. (This is not to be confused with the 29 day period the Moon takes to complete a revolution around the earth.) Hence this 27 fold division is a very natural one, and as a result it naturally also plays a pivotal role in Vedic astrology.

We have seen that for civilisations which last through many millennia, the wobble of the earth's axis of rotation becomes relevant for time-keeping. This has a period of about 24,000 years, and it causes a precession of the zodiac relative to the earth and the Sun. This means that over long periods of time such as millennia, the Sun's position at a particular time of year will move backwards through the constellations of the zodiac. Given that there are 12 signs in the zodiac for the Sun and the period of the wobble of the earth's axis is about 24,000 years, the Sun effectively shifts one sign position every 2,000 years. The direction of the wobble appears to make the Sun go back one sign position every 2,000 years. For example the Sun was entering the beginning of Aries on 21st March around 500 CE, but now (over 1,500 years later) it does not in fact enter Aries till around 14th April. The Sun's position at 21st March has gone back a sign into Pisces. This apparent precession of the Sun through the zodiac is relevant for time-keeping across several millennia as it gives a very simple way of indicating a millennium using a measure (the stars) that is universal and fixed for many millions of years. This is in contrast to the measure we have today, which is arbitrary to the prevailing culture. Hence it is easily prone to change, and is subject to a loss in accuracy in just a few centuries. It is a reflection of the shorter term cycles that modern civilisation is focussed on, which will also naturally tend to give it a shorter life.

The precession gives us important information with regard to the dating of the Rig Veda and Vedic astronomy. They are several passages in the text of the Rig Veda referring to the position of the equinoxes in the zodiac.[3] One of them has the spring equinox (21st March) in the zodiac sign of Cancer.[4] Today we see that it is occurring in Pisces, which is four signs behind Cancer. As the Sun takes about 2,000 years to go back one sign, it would not be unreasonable to conclude that that particular reference is to a period of time around 8,000 years ago. The seeming eternal longevity of the Vedic culture, which survives till today, is due to the emphasis it places on the real and the eternal. To this extent one of the main purposes of this book is to show that its philosophy and approach to life is a natural and eternal one.

Vedic astronomy and astrology have been highly developed for many millennia now, as is reflected in ancient texts such as the Vedanga Jyotisha and Surya Siddhanta. These are compilations of information from even earlier times and whose origins are uncertain. The dates of the compilations have been conservatively estimated at being around 2,000 years ago according to modern thought. Apart from the knowledge of the precession of the Sun, these texts give astonishingly accurate positions of the planets and other heavenly bodies, especially when we assume that they were calculated without telescopes. The precession is one of the important factors that distinguishes the Vedic system of astrology from 'Western' astrology, where it appears not to have been known. It means that Western astrology is astronomically inaccurate. For example, Western astrology considers the Sun to enter Aries on 21st March, but as we saw, it is currently around 14th April.

The Four Yugas

We can observe that cycles of time are generally divided into active and passive periods. For example the 24 hours of the day are divided into the lighter hours of work and the darker hours of sleep. Similarly the 12 months of the year are divided into the warmer and lighter summers, and the darker and cooler winters. The Rishis of the Vedas also saw such divisions of periods for the longer cycles of times that the Vedic culture was able to experience. These longer divisions are called yugas or 'great ages'. In these longer time frames the Rishis were able to notice that not only physical activity, but that also the mental activity of humans varies between an active and a passive phase. In particular they were able to notice that the level of direct perception varies according to whether it is the darker or lighter period or yuga.

There are different definitions of yugas for different time cycles, but here we will take a look at the one associated with the 24,000 year cycle of the earth's axis already mentioned, as it is most relevant to the frame of time that is being considered in this book. The modern statement of this yuga system is attributable to Sri Yukteshwar, the guru of Yogananda Paramahansa, in his book The Holy Science. This divides the 24,000 year period of the Sun's precession through the entire zodiac into four yugas, known as gold, silver, bronze and iron. The darkest point is in the iron yuga, and this is said to have last occurred when the Sun entered Aries on 21st March, i.e. around 500 CE. It is in the lead up to this period that we can observe the collapse of great cultures beginning with the Egyptian, South American and Chinese Tao cultures, and later with the Greek and Roman cultures. In Europe this period ushers in what are commonly known as the 'dark ages'. In India we can similarly observe a parallel shift in the Vedic culture towards the direction of a superficial ritualism, though it did manage to survive through that period.

The lightest point (gold yuga) is therefore 12,000 away from 500 CE, i.e. 11,500 BCE the last time around, or approximately 10,500 years from now the next time around. According to this system of the yugas, it was around the last period of maximum light that the Vedas as we know them today were seeded. It was a time of light as far as direct perception is concerned, but of darkness as far as physical or sensory perception was concerned. Hence ideas of advancement and development were very different from today where we have only moved a little away from the darkest point of direct perception, and the lightness and darkness of our perceptions is almost reversed from those ancient times.

Some Basic Principles of Vedic Astrology

Vedic astrology is based on the 12 fold and 27 fold zodiacs which are rooted in the Rig Veda. We have seen how the position of the stars and planets at our birth (the natal chart or horoscope) is like a karmic map for us. The natal chart uses the 12 fold division of the zodiac to represent 12 domains of life (the houses), and the light of the stars (constellations), planets, Sun and Moon present in these houses form the framework for our lives via the connection between light, time and karma. The basic principle of Vedic astrology rests on an examination of the houses, with the constellations influencing the nature of the houses as background fields, and the Sun, Moon and planets influencing the houses more as focussed energies.

The region that is rising in the east at the time of birth is of particular importance for the natal chart. This indicates the type of energy that is increasing in influence at that particular time, as light and energy from that area is coming into prominence. It is much like how the rising of the Sun in the east influences the beginning of a new day. This region is called the first house (or the ascendant), and it governs the key outer characteristics of a person. As such it is the most important house from the perspective of daily life. Similarly, the house opposite the first house is the one that is descending in the west, and it represents the light that is reducing in influence. It is like the complementary dual of the first house, and so it represents inner areas more such as relationships. As it is opposite the first house, it is six houses away and it is known as the seventh house.

Another region of importance is the one that is vertically above (mid-heaven) at the time of birth. It represents the area of maximum activity, and it thereby relates to the main work and outer activity of a person. It is in between the first and seventh houses, and so it is the tenth house in the circle of 12 houses. Directly opposite it is the fourth house, and this represents the complement of outer activity, or the inner activity of a person. It thus signifies the home and emotional life. As these four houses are usually the key domains of life, they are typically given prominence in Vedic astrology charts, both in terms of representation, and in terms of interpretation. It is possible to extract meanings similarly for the rest of the eight houses, and for those who are interested in this further there are many good books available on the subject,[5] some of which have been listed in the bibliography.

Apart from the houses and the 12 fold division of the zodiac relating to the Sun, the role of the Moon is also given great importance in Vedic astrology. It brings in the system of 27 constellations, and they help to give additional insights into the interpretation of the chart. As the Sun symbolises an energetic light and the will, so the Moon symbolises a more cooling and gentle light. As a reflection of Sun light, the properties of Moon light are also reversed. Therefore it is what brings us physical and mental stability, and it is perhaps the most important factor in the emotional life. Due to the importance of emotions in daily life for us, it is the role of the Moon that makes Vedic astrology relatively accurate. In addition to the knowledge of the precession of the Sun, lunar based astrology is also not present in Western or Greek based astrology.

As the Sun and the Moon are particularly dominant sources of light on earth, two extra ways of looking at their relationship are considered. They are the notional planets of Rahu and Ketu, and geometrically they are represented by the intersection points of the earth's orbit with the Moon's orbit. Rahu is the intersection via the North Pole of the Moon and is associated with lunar eclipses, and Ketu is the intersection via the South Pole of the Moon and is associated with solar eclipses. As eclipses markers they particularly represent the inner or mental Moon and Sun light.

The planets used in Vedic astrology form the pentuple of Mercury, Venus, Mars, Jupiter and Saturn. To look at their meanings and implications would once again require a more detailed study of astrology. Together with the Sun and the Moon, they represent the complete septuple described in the chapter on Cosmology. The Sun and the Moon can also be considered as the stellar equivalent of Purusha and Prakriti, and with the group of the five planets, they form the stellar equivalent of the cosmic triad. In this case the pentuple of the planets relate to the five elements that emerge from the cosmic triad.

Significant Days

In the Vedic system significant or holy days are also determined via astrology and astronomy. As far as the year is concerned, the days when the Sun light begins to increase or decrease are considered important for the same type of reasons that the four main houses in the natal chart are considered important. These are the solstice points of the year, and the winter solstice is particularly important as it represents the start of new energy and activity for the whole year, much like the first house in the natal chart. The equinox points represent balance between darkness and light, and are important as transition points from inward to outward orientation, and vice-versa. This type of reasoning can be applied to the daily cycle as well, with the sunrise, sunset, midday and midnight all being similarly significant.

The position of Moon also determines significant days. The full Moon days are supportive and nurturing, whereas the dark Moon days are harsh and so are useful in developing detachment. There are several other significant days annually, and typically they can be quite complex to calculate (they bring in the constellations too), but the main reasons by which they are determined have been indicated here.

Ayurvedic Astrology

We saw that the Vedic sciences are not discrete disciplines, but are a part of one unified system as they are based on the common underlying principles of Vedic philosophy. Therefore it is natural for the different disciplines within the Vedic system to overlap and flow into each other. To this extent we have already noted some of the connections between Yoga and Ayurveda, and Yoga and Vedic astrology. Now we can take a look at some of the connections between Ayurveda and Vedic astrology too.

Vedic astrology gives us guidance as to an appropriate way to pursue our lives on an individual level, or guidance with regard to our individual dharma. When we apply this at the physical and instinctual levels this dharmic guidance is the harmonious life according to Ayurveda. This is the principle of maintaining our natural dosha balance. By the principles of Vedic cosmology and the five elements, the stars, Sun, Moon and planets must have doshas too according to their composition by the elements. From their influence on the houses of the natal chart we can get further insight into our doshas and any potential disturbances.

As an illustration, the doshas of the Sun, Moon and the planets used in Vedic astrology are indicated below:

Vata – Saturn, Rahu and Mercury
Pitta – Sun, Ketu and Mars
Kapha – Moon, Venus and Jupiter

The rationale for some of these correlations is obvious, such as the connection of the Sun with Pitta dosha. On an examination of the archetypes represented by the planets the other correlations become clear too, but this would require a level of further study inappropriate to the purpose of this book.

As each of the doshas is represented by three heavenly bodies, Vedic astrology gives us an extra level of insight into the nature of the doshas involved in us, including at the mental levels. Additionally, due to the snapshot nature of the natal chart, we get an insight into the strength of a person's constitution, irrespective of their current condition. Its cyclical nature also indicates stronger and weaker times in advance, which allows for suitable preparation well beforehand. Thus it is possible to fulfill karma harmoniously, or dharmically, rather than by accident. This makes life a more positive learning experience and minimises unnecessary suffering and illnesses.

Astrological Therapies

Dharmic living, or living according to karma, forms the basis of astrological therapies in the Vedic system. In addition to the Ayurvedic approach to this area, there are certain therapies that specifically relate to Vedic astrology. Light or colour therapy is an important example of this. As astrology is primarily an examination of light this form of therapy naturally relates to astrology. We know that there are seven colours in the visible spectrum, and these relate to the seven major heavenly bodies. Additionally the invisible planets of Rahu and Ketu relate to the invisible part of the spectrum, or the ultra-violet and infra-red parts respectively. The main way this therapy is practised is via gemstones and crystals of colours that match the heavenly bodies.

Another important therapy in Vedic astrology is sound or mantra therapy. We saw how sound and light are particularly connected with each other as Devas and mantras, and the light of the heavenly bodies thus naturally relate to mantras too. There are mantras that relate to the heavenly bodies directly, and others that relate to a particular heavenly body via its dosha. The mantras can be of the bija form already looked at, or longer Vedic verses. As such mantra therapy is Deva therapy, and it is the highest dharmic therapy in Vedic astrology.

A Precaution

It is important to note that we should take the advice of a skilled Vedic astrologer when we consider gemstones, mantras and Devas for the purposes of astrological therapies.

ENDNOTES

1. In particular, the hymns 1.140 to 1.164 of Rishi Dhirgatamah of the Rig Veda. See Gods, Sages and Kings by David Frawley for more on this.
2. As above.
3. As above.
4. Rig Veda 10.19.1, and as above.
5. For example The Astrology of the Seers by David Frawley, and Elements of Vedic Astrology by KS Charak.

8.

Yoga:
The Science of Unity of Mind

Lead me from the Unreal to the Real.

Brihadaranyaka Upanishad 1.3.28

The Birth of Yoga

Yoga means union, and its primary goal is to facilitate the union of the individual with the universal. This makes it the highest Vedic science, and therefore it reflects the entire purpose of the Vedic system of thought. It is the primary means of the fulfilment of the evolutionary impulse to the stability of unity, and as a result it represents all forms of dharma. As such we can consider any of the Vedic disciplines as a path of Yoga, and its vast scope is evident in the many variations and approaches to it that exist. Ultimately Yoga achieves unity by the power of the one-pointed concentration of the mind as a state of pure consciousness or reality.

The birth or origin of Yoga is the same as that of all manifestation or Prakriti, for the concept of a union takes meaning as soon as duality is established. The question of origin or creation (and destruction) of manifestation (and time) is not something that happens in time, but from a transcendental state into and out of time. It is a reflection of the state of the subject either as the transcendental sakshi (Purusha as subject and object) only, or the individualised consciousness as mind (the subject) that exists in time along with manifestation (Prakriti and the triads/pentuples as the objects). As soon as there is the concept of time and manifestation we have the law of universal karma. At the same time we must also have universal dharma, for the universal karma implies that what comes into manifestation and time must also come to an end, or go out of time and become transcendent. The way by which manifestation becomes transcendent (Purusha) is dharma, and this is facilitated by the paths of Yoga. As all manifestation is said to be born via the golden womb or Hiranyagarbha, the birth of Yoga is also attributed to Hiranyagarbha.

Yoga as Sanatana Dharma

Thus Yoga is the original and eternal dharma, which is known as sanatana dharma in Sanskrit. Sanatana dharma is the basis of all Vedic activity and life as it the first natural impulse to action that occurs in the world of time. Generally it is concerned with any action that naturally facilitates the evolutionary impulse, and it functions within the framework of the law of karma. As a result it is naturally very inclusive, and it does not tie its self down to any particular set of texts, teachers or prophets. It is not only one way or one path, but it represents a combination of many such approaches according to the need of the time or the individual.

In later times the term Hinduism came to be known for the Vedic tradition of sanatana dharma, largely due to influences external to the tradition. It was part of an attempt by the later religions from the Abrahamic tradition (Christianity, Islam and Judaism) to categorise the Vedic tradition in its terms. In fact, an element in common with the Abrahamic approach does exist within the Vedic tradition as a part of it, for it also has its place in the natural order of things. However the purpose of the Vedic tradition causes it to bring together complementary dualities such as subjects and objects, or arts and sciences, and it similarly does the same with the duality of scientific rationality and religion too. This is where the Vedic tradition distinguishes itself from the Abrahamic and other traditions, and we have seen that rationality itself demands this type of approach.

To this extent a superficial examination of sanatana dharma can lead to the very misleading conclusions which are evident in modern scholarship.[1] Modern thought struggles to see the underlying connection of the vast range of its activi'ties, and therefore it does not see the inner or subjective aspect of its outer rituals. It is a naturally difficult tradition to categorise, because it encompasses the vast breadth of nature, and nature cannot only be comprehended by analytical thought as we have seen through the course of the book. It also requires the deeper and more natural form of thinking of direct perception.

Viveka, Vairagya and the Three Main Paths of Yoga

Two complementary concepts central to the development of the various paths of Yoga are known as viveka and vairagya in Sanskrit. They naturally arise from the unified equilibrium of reality, and they relate to the Satchitananda (being, consciousness and bliss) of Purusha. Viveka is the faculty of discrimination that we have, and it includes rationality and direct perception. It is the ability to discern the path to the real, or to discriminate between that which facilitates evolution from that which takes us away from the real. It is the natural result of the urge to the stability of a unity. It enables us to see reality in all, initially by the path of rationality, and later by the way of direct experience. It is therefore a wisdom approach, and it culminates in a state where there is only reality as knowledge. It relates to Chit (seed of intelligence) of Purusha, and gives rise to the path of wisdom or Jnana Yoga.

Vairagya is the complementary idea of detachment in the Vedic system. It is the ability to detach from the unreal and to exist only as the real. The concept of detachment is contained in the unity of reality, as attachment requires more than a unity. There needs to a subject that gets attached and an object of attachment. Subsequently a third aspect, which is the process of the attachment between the subject and object, is also required. If we are in reality a unity, then detachment is a natural process as it takes us to the natural state of being a unity only.

The real power of Vairagya lies in its ability to overcome the problem of self-reference. It detaches the subject from the object, and as a result it gives the ability to make true observations. When we are attached to or a part of something, we cannot really observe it for reasons of self-reference as explained. Vairagya provides the fulcrum to see things as they really are, and as such it demonstrates an important universal principle that to gain mastery over something also requires detachment from it. For instance to realise true happiness, we must detach from the idea of happiness and unhappiness. To understand the law of karma, i.e. the workings of the law of cause and effect and time, requires detachment from the idea of karma or us as the idea of the doer. We must stand outside of manifestation to truly observe it deepest laws via the direct perception of the sakshi.

Thus Vairagya relates to the Sat (being or existence) aspect of Purusha, and it gives rise to the path of Karma Yoga where we detach from thinking of ourselves as the doer. Ultimately vairagya goes further than detachment, for detachment implies the existence of attachment as a complementary dual. Instead it leads us to the state of non-attachment, which goes beyond even subjects and objects or all dualities such as attachment and detachment.

We can see that there is an urge towards the equilibrium of Purusha underlying the above two paths of Yoga. It is the universal dharma due to the law of karma. This urge is the basis of the third main path of Yoga, Bhakta (devotional desire) Yoga, where we devote or surrender ourselves to our natural desire for unity. When this desire is fulfilled or transcended into a unity we achieve peace and contentment, and so this path of Yoga relates to the Ananda aspect of Purusha. These three paths of Yoga naturally reflect the concept of the triad in nature. In fact they are not thee mutually exclusive paths, but ultimately they are one as is reflected by the unity of Purusha as Sat = Chit = Ananda. In order for Yoga to be successful all three paths are required. For without a foundation in Bhakta, Yoga cannot commence, without viveka we would not know what to detach from, and without vairagya we would not be able to discriminate the real from the unreal. However it is possible for one of the paths of Yoga to dominate according to our individual nature, as we shall see.

These three paths of Yoga enable us to overcome our seeming ignorance (avidya) of Purusha due to Maya, where we mistaken a temporary triple stability for the real stability of unity. However, while we think we are in a three-fold world we are able to make use of the tools of the triple world of nature to achieve real stability. Therefore we have the three paths of Yoga above as well as the other Vedic sciences. They are the means of detaching from dirty fuels, or discriminating between clean and dirty fuels. To this extent we can relate the paths of Yoga to the doshas of Ayurveda too. Kapha's nurturing qualities relate to the devotional path of Bhakta Yoga. Pitta's ability for insight enables us to see or discriminate, and so it relates to Jnana Yoga. Vata has the motion of upliftment or detachment which relates to Karma Yoga. Obviously these divisions are not rigid, and there must be overlaps as with all of nature.

Patanjali's Integral (Raja) Yoga

To reflect the underlying unity of the three main paths of Yoga, a unified or integrated system was established known as the Integral or Raja (royal) Yoga. Its main exposition is a compilation by the sage Patanjali over 2,000 years ago (the dating of the knowledge it contains is unknown), in the text known as The Yoga Sutras of Patanjali. This text forms the basis of the classical system of Yoga philosophy, and it is one of the six main philosophies of the Vedic system discussed in the first chapter. The Yoga Sutras consist of 200 statements or sutras (which literally means threads), and they combine all the three paths of Yoga. They ultimately aim for the mental state of one-pointed concentration that results in the knowledge of the unity of pure consciousness, or the reality of Purusha.

Patanjali's Yoga consist of eight major limbs, which is known as ashtanga in Sanskrit, and therefore it is sometimes referred to as Ashtanga Yoga too. The eight limbs in order are:

1. Yama
2. Niyama
3. Asana
4. Pranayama
5. Pratyahara
6. Dharana
7. Dhyana
8. Samadhi

In this section we will take a brief look at the meanings of these eight limbs. The first five are known as the outer limbs and the last three as the inner limbs, for reasons we shall see.

The yamas and niyamas are each five recommendations for conduct that we should try and follow in our daily lives. These pentuples are not arbitrary, but are a reflection of the universal five fold nature. The Yoga Sutras begin with these as they concern our emotions and intellect, which is the main areas of activity for most people. The yamas are restraints on behaviour:

1. Non-harming (Ahimsa)
2. Truthfulness
3. Non-covetousness
4. Abstinence (from things that take us away from the real, i.e. dirty fuels)
5. Non-possessiveness

In the yamas we can see the basis of the path of Karma Yoga described, as they form a foundation for outer detachment. The niyamas by contrast are observances we should try and maintain:

1. Surrender to Ishvara (to the Devas, or the Divine)
2. Self Study
3. Will or Effort to Activity (Tapasya)
4. Purity
5. Contentment

In addition to Karma Yoga the niyamas also form the basis of the path of Bhakta Yoga. The codes of conduct are not meant to apply just to the outer life, but should also be the aims of our inner lives. For example abstinence or non-covetousness are not just physical restraints to consumption, but ideally they also should reflect a lack of mental desire for an outer life. Hence they are not about repressing desires for an outer life, but developing a desire for the real instead.

The next limb is asana. In the modern practice of Yoga a lot of emphasis has been put on asana as form of physical exercise. The word asana only occurs in two or three sutras of the 200 sutras in the Yoga Sutras, and it is generally taken to mean a comfortable physical posture. As such it is not about a perfect body, but the ability to forget the body. This is important for the subsequent limbs of Yoga as it does not allow the distracting thoughts of an uncomfortable body to become an impediment to achieving a state of one-pointed concentration. Today the physical nature of society has put emphasis on the aspect of asanas relating to the Vedic system of physical exercises. In fact most of the practice of Yoga has been reduced to that. Physical exercise is a natural and important way to still the body and begin the practice of the Yoga of stilling the mind, but its role and place within the tradition should be recognised. It falls more appropriately under Ayurveda as it primarily concerns the physical body.

The fourth limb of Patanjali's Yoga is Pranayama. After attaining a comfortable posture the next thing to gain control and comfort over are the instincts or Prana as the life-force, and so the meaning of Pranayama is control over Prana. As the breath is the most important instinct we have to receive Prana at the physical and instinctual levels, the breath is also the best way to facilitate Pranayama at these levels. Though specific breathing exercises are not mentioned in Patanjali's sutras, several breathing exercises are included in this system of Yoga for this purpose. In the subsequent limbs of Patanjali's Yoga Pranayama is achieved via the force of the mind itself.

The fifth limb is Pratyahara (withdrawal of the senses) where the emphasis is on the senses and the mind. After the instincts have been calmed the next level that needs to be worked on is the level of the senses. For this purpose the important realisation that the senses are not just stimulated by the external world, but also by the internal world, is cultivated. Thus the senses are trained to detach or withdraw from the outer world, and to focus instead on the inner world of the tanmatras. Pratyahara is a form of vairagya, and it serves as a bridge from the five outer limbs of Yoga which mainly address the lower koshas, to the three inner limbs which mainly address the higher koshas.

The three inner limbs of Yoga aim to achieve the unity of reality by developing a state of mind that can reflect it. This requires a mind that has the calmness and unity of reality, or a mind in a one-pointed state. It is a mind that can control the variety of manifestation that occurs in it, and so the inner limbs of Yoga represent the real asanas as the asanas of the mind. They enable the direct perception of the higher dosha functions as the stilled mind is able to just see or reflect, without the disturbances of external or analytical thoughts. This is the basis of aiming for Sattva in Yoga practices, and progress towards a calmed mind is progress through the various stages of Sattva via the evolutionary power of Agni within us. While Agni facilitates the metabolism of progress, it has a Rajasic element of activity, and it relates to tapasya in the niyamas above. Once evolution is complete, Agni is evolution or knowledge itself within us.

The first of the three inner limbs is Dharana, or the power of concentration. Here Sattva is developed by concentrating and gaining control over the internalised mind of Pratyahara, rather than being led by the tanmatras and the thoughts generated by it. Various concentration techniques are employed for this. First amongst these are mantras,[2] and the mind is trained to be immersed in just the one thought of the particular mantra being used. In the chapter on mantras we saw that there were three different levels of speech, i.e. audible (vaikhari), whispered (madhyama) and silent (pashyanti). Therefore according to level of practice of the mantra, it is has the power to take us directly to the highest Yogic state of samadhi. Other concentration techniques include visualisations such as on the elements of nature like the sky for space, the clouds for air, a ghee flame for fire, a lake for water and the mountains for earth. Concentration techniques are ways of improving the quality of the fuels (Prana and Soma) we take in at the mental levels.

The second inner limb of Yoga is dhyana, or the state of meditation. Dhyana is the state of sustained and one-pointed concentration on the object by the subject, and thus it is where the ability to directly perceive is developed. It lifts the doshas to their higher functions, and this allows for the advanced practice of viveka and vairagya to become established. Pitta as Agni enables us to see or reveal the real from the unreal. Vata as Vayu detaches or eliminates for us the unreal, and uplifts us to the real. Kapha as Soma gives us the strength and desire in order to develop viveka and vairagya.

As there are different techniques for dharana, there are also different approaches to meditation. A Pitta approach begins with rationality, which is also the starting point of the Vedantic approach, as we have seen. Therefore this approach to mediation commonly begins by asking the question 'who am I?'. In the limited rational phase of this meditation it can only proceed negatively with the question 'who am I not?', as rationality does not have the tools to address open or infinite questions. When through this analysis the thought 'I am' only remains, and there is complete viveka, direct perception can dominate rationality. Vedantic meditation then can reveal more positively that instead of 'who am I not?', in fact 'I am all'. It is its culmination in samadhi, which is the final limb of Yoga as described below. This technique of meditation was emphasised by Ramana Maharshi, perhaps the greatest Jnana Yogi of India over the last few centuries.

The Kapha approach to meditation is one based on desire, and so it is more devotional in nature. It involves offering ourselves to a particular divine quality or Deva. The Vata approach is more variable then the other doshas which reflects its airy nature. It involves detaching from the unstable, and coming to a rest in the stable. This creates the space for the possibility of direct perception to manifest as a flash of inspiration.

The final limb of Patanjali's Yoga is samadhi. This is the culmination of Yoga, and it is the state of pure direct perception where the subject and the object are completely merged and manifestation is transcended. As a result there is no concentration, for there is no subject to concentrate, nor is there any object to concentrate upon. Direct perception goes beyond concentration, and there is just the awareness of the existence of the unified subject and object in this state. The unified awareness of the subject and object is like a causal seed of the pure creative power of a causal mantra or a Deva, and so such samadhis are called 'with thought or seed' (savikalpa). When there is not even the awareness of the unified subject and object, but a samadhi on the unity of transcendence, the samadhi is called 'without thought or seed' (nirvikalpa). As this second type of samadhi is transcendental it is the same as Purusha.

Samadhi and Models of Reality in Manifestation

We can thus see that samadhi is a complete stillness of mind, and as result during samadhi there can be no manifestation via the mental activity of consciousness. Instead there is only the reality of the causal world or transcendence. However rationality requires that ultimately manifestation must be based on the underlying causal reality, and we can see how this can come to be via the role of samadhi.

Manifestation presents itself to us in all its infinite variety as an infinite number of different flashes of reality, with each flash revealing different aspects of the seeds in the infinite reality.[3] This is in contrast to the sustained and real state of Purusha. By stringing the flashes together, we create the picture of time and causation that is manifestation. The flashes are equivalent to a series of 'fleeting' samadhis,[4] as opposed to a sustained samadhi. Each aspect or moment of manifestation requires a samadhi. To realise the existence of a flower requires a fleeting samadhi between the observer and the existence of a flower. The recognition of its colour requires another such samadhi.

As any samadhi is completely one pointed we must lose consciousness of our personal self for the moment of even a fleeting samadhi. We can observe this by the fact that when we are completely engrossed in something (samadhi), we lose consciousness of our self. For example, when we witness a gruesome act we become unaware of anything else, including ourselves. We only become aware of ourselves when the object of our fleeting samadhi is ourselves. As our personality is usually the most lingering thought and awareness we have, it is also the most frequent fleeting samadhi that we have.

Usually samadhis are described as exhilarating experiences as they put us in contact with the real. We experience the bliss of life as Soma which is Ananda of Purusha. We forget the observations of the personality, and instead we have the direct perception as the sakshi or

Purusha. The extent this experience manifests in us depends on the type of samadhi we have. Recognising a colour of something at a particular moment may not as blissful as the inspiration of a poem, though even such basic sensations can be exhilarating if we can capture the joy (Soma) that is present in all life at that moment. This is the principle that underlies Tantra as we shall see in the next chapter.

Fleeting samadhis are also a way of understanding the connection of the physical world to the causal world of the original doers or original manifesters. They show us how physical manifestation goes back to the causal level of Ishvara, and they also show us how it is really the Deva kingdom within us that underlie all levels of manifestation and evolution. Thus the real seers and hearers are the Devas, and this gives rise to the Upanishadic statements such as 'the eye of the eye' and 'the ear of the ear'.[5] When we confuse the Devic actions of manifestation and evolution with each other due to the blind use of emotions and the intellect, we make the archetype of the Devas into the archetypes of the demons within us. This is a result of confusing manifestation with reality, or having an unstilled mind of fleeting samadhis as opposed to the perfectly stilled mind of a sustained samadhi. Sustained samadhis overcome the blind mind by giving the 'sight' of direct perception.

Our samadhis also have the power to define our concept of reality, whether as individuals or as a society. Fleeting samadhis at the level of the physical world give us a material view of reality. Samadhis on the emotions and the intellect make us experience reality at that level. The more sustained a samadhi, the more it takes us towards the universal and eternal reality of Purusha. As a result we look for the deeper meaning of our lives and make our religions according to the level of our samadhis, which is the level of our understanding of reality.

In emotionally dominant times the framework for emotional religions is formed as we saw in medieval Europe.

After that when rationality began to dominate in Europe, we saw modern science fulfill the role of a religion. More recently there has been a trend for the sensory kosha to dominate across the world, and this has lead to the religion of world-wide consumerism. In Vedic times we had the religion of a universal reality based on the direct perception of sustained samadhis. A present we can see the clash of all of these models of reality, and so the eternal battle repeats itself as depicted in the great 'eternal' texts of history (itihasas), the Mahabharata and the Ramayana.

The Three Gunas and the Four Castes

Apart from the importance of the role of Sattva guna in Yoga, the three gunas together also form the framework for the division of humanity into four different types or castes. The purely Sattvic type is of a highly balanced or spiritual nature, and is the basis of the brahmin caste. Where there is Rajas moving towards Sattva via the evolutionary action of Agni, we have the fire of the battle towards to reality. This underlies the warrior caste of the kshatriyas. Where the action of Agni is not part of an upward oriented triangle, we have Rajas moving towards the materiality of Tamas. Here activity concentrates on material accumulation and we have the basis for the business caste of the vaishyas. Finally where there is pure Tamas only, we have the physical dominated humanity of the shudra caste. From this we can see that every one of us must reflect all four of the castes, but in different proportions according to our nature.

Thus caste is determined by our qualities (the gunas above) and also by our actions (karma), and not just by birth. As such the caste divisions are universal, and not a property of any particular system. We can see that even some great sages of Vedic times would be of the lowest caste, if they were judged by their birth[6] alone. Caste has been particularly misapplied in India during the iron yuga, and this has lead to the weakening of its society by distancing it from its Vedic heritage. Any such significant and unnatural division of society is bound to harm it, and in this case it led to the easy foreign occupations of India, which caused the further demise of Vedic culture in the land.

Other Paths of Yoga

In addition to the systems of Yoga that have been already outlined, there is another important path of Yoga known as Tantra. It focuses on the force and energy (Prana and Shakti) at the level of all the koshas, and it is more technical nature. By contrast the Yoga systems described earlier in the chapter focus mainly on the will, or the Prana of the mental koshas. The energetics of the lower koshas is implied in them. Some examples of the techniques of Tantra are the systems of Kundalini Yoga and Hatha Yoga. As these energetic Yogas have a stronger focus on the lower koshas they are usually more easy to access, and they form the subject of the next chapter.

In the final chapter of the book we will also see the Vedic Yoga of the Rig Veda. It is the Yoga of Yajna or the concept of sacrifice at a multidimensional or metaphoric level, much like the language of the Rig Veda. We should note that eventually the methods of all the paths of Yoga must unite by the virtue of the definition of Yoga as the science of union.

ENDNOTES

1. One needs to only look at the definition of Hinduism in the various encyclopaedias of today. For an excellent and in-depth, but yet easily accessible overview, refer to Ways to Truth: A View of Hindu Tradition, by Ananda Wood.
2. For a guide to mantras, see Mantra Yoga and Primal Sound by David Frawley.
3. This Yogic model of reality is startlingly similar to the conclusions being forced upon modern thought by the field of Quantum Mechanics.
4. This is explained in the Tripura Rahasya.
5. Kena Upanishad 1.2
6. For instance, the story of Satyakama Jabala in the Chandogya Upanishad 4.4.

9.

Tantra: The Science of Energetics

Express the Nectar of your Wisdom, Goddess, the perfect Word in God.

Rig Veda 7.34.9

The Origins of Tantra and the Rig Veda

Tantra means technique or an instrument of expansion, and it also means to weave due to the way the technique of weaving expands a thread into a piece of cloth. As a reflection of this symbol, Tantra uses techniques of energy and force (Shakti and Prana) to weave us into the fabric of the entire universe, or to expand us from the individual to the universal. To this extent it aims to see the reality that underlies all the energy of nature, and it even uses its techniques in the activities of daily life. It eventually seeks the bliss (Soma) of perfect rest from expending energy, which is the bliss of the completeness of reality. In contrast to Raja Yoga where mental control is explicit, in Tantra mental control is a consequence of energetic control.

Tantra exists so long as there is energy, and so like any system of Yoga its principles are a part of the fabric of nature. As far as the usage of Tantra in human history is concerned, we can see metaphorical references to Tantra in the Rig Veda. For example, one of the common images portrayed in the Rig Veda is Indra drinking the nectar of Soma brought by the action of Agni.[1] Indra is a metaphor for the divine mind, and so this suggests how energy can be a source of bliss for even the deepest levels of consciousness. One of the great practitioners of Tantra of modern India was Ganapati Muni (the chief disciple of Ramana Maharshi), who showed many connections between Tantra and the Rig Veda.[2]

Over the last two thousand years or so Tantra has been presented in a more explicit form than in the Rig Veda, in a set of texts known as the Agamas. There are different branches of Tantra within the Agamas, but they principally relate to the deities of Shiva and Shakti as a dual, and also Vishnu. Perhaps the foremost of these branches is known as Kashmiri Shaivism, and its premier exponent was Abhinavagupta who was another one of the great Tantrics of India from about the 10th Century CE. In particular, he showed how the Soma of reality relates to the inspiration and delight of artistic expression.[3]

Hatha Yoga and Tantra

Hatha Yoga is an energetic or Tantric system of Yoga, where the metaphorical Agni and Soma of the Rig Veda are given more concrete definitions. Agni here refers to the fire or energy at the base of the spine, or the kundalini shakti. Soma is associated with the crown chakra at the top of the head, or the bliss kosha. The aim of Hatha Yoga is for Agni to rise up the spine and melt the nectar of Soma at the top of the head, which reflects the metaphor from the Rig Veda above. The Soma is drunk to the satisfaction of the divine mind Indra, and this has the effect of awakening Indra within us.

As the main focus of Tantra is energy, it accesses this at the level of all the koshas. Hence Tantra focuses more on the physical body and instincts than the other systems of Yoga. We see that physical asanas are particularly well developed in its system. The emphasis on the physical body today has brought asanas to the forefront of all the systems of Yoga however.

A Comparison of the Rig Veda, Yoga, Vedanta and Tantra

We have already seen how Yoga, Vedic Astrology and

Ayurveda relate to each other. Similarly, it is now also helpful to see how the Rig Veda, Yoga, Vedanta and Tantra all relate to each other, in addition to the metaphorical references to Tantra just seen in the Rig Veda.

Raja Yoga aims to realise reality through samadhi, primarily by the force of the will. This is a direct control of the Prana of the mind and the higher koshas. Tantra by contrast, aims to realise unity by a control of the life-force from the level of the lower koshas upwards. This involves a control of the Prana from the instincts onwards, and as a result the psycho-physiological energy or Shakti at all levels in the person. Mental control is an implication of this multi-layered approach.

The Rig Veda aims at an inclusive and metaphorical description of reality. It is the experiential approach of direct perception. The Vedantic Upanishads are initially relatively more intellectually oriented, and hence by the limitations of rationality, Vedanta is less able to give an inclusive description of the infinite nature of reality in its intellectual sections. Instead it has exclusive descriptions by stating that reality cannot be anything that can be conceived rationally. Thus the Vedantic approach is able to comprehend Maya, and by which it is possible to negate all of manifestation as illusory or unreal.

Like the Upanishads, Tantra aims to bring reality into a lower domain of manifestation than the Rig Veda. However it aims for an inclusive approach, which is where it differentiates itself from Vedanta. In order to do this it aims to experience the bliss of reality in manifestation not by an intellectual approach, but by uniting with the deepest essence of manifestation via the Prana and Shakti present in it. The methodology behind this will be the main focus of this chapter.

Notice that ultimately all of these approaches must merge into one in reality. For instance Raja Yoga is both an inclusive and exclusive approach, because by concentrating the mind it excludes manifestation, but it becomes all inclusive in samadhi. By negating manifestation Vedanta ultimately directly perceives manifestation as only one reality. Tantra eventually negates manifestation when it experiences the bliss of unity. We could say that the Rig Veda has the highest approach as it aims to directly perceive reality without the intermediary techniques of the other approaches. It relies more on the higher faculties of inspiration, revelation and intuition, rather than the preceding stages that the others rely on. However it contains these other approaches in seed or metaphoric form, as we have seen through the course of this book.

Thus each of these approaches must overlap with each other, and in fact each approach can be viewed as containing the others within it. The approach we choose to begin with depends on our doshas and gunas (qualities), and they in turn depend on our individual dharma and karma.

Shiva and Shakti, the Tantric Couple

In this chapter we will focus on the system of Tantra that employs the symbolism of Shiva and Shakti, and which is perhaps the main Tantric system. In the chapter on Cosmology we saw that the force of Prana arose between Purusha and Prakriti, thereby establishing this great dual as the divine subject and object. Tantra focuses on the resultant energy or Shakti that arises in this triad. Here Purusha and Prakriti are related to Shiva and Shakti, with Shakti representing the energy of Prakriti arising due to Prana. It is this energy that forms the stability of the triad in place of the real stability of a unity. The main purpose of Tantra is to cause the dual of Shiva and Shakti to unite into the bliss of one stable reality via the work of the energy of Shakti. This eventually involves Shakti removing itself as a factor of the division of the unity, or by Shakti realising its unity with Shiva. The final unified state is also known as Shiva, but in his form as Purusha. Thus we can see that Tantra specifically deals with the unification of the divine couple of Shiva and Shakti (at all their levels) via energy.

Before the unification of this transcendent triad can take place, there are also the many lower triads representing the various levels of manifestation that have to be similarly unified. Any unification of a complementary couple actually involves the concept of a triad, whether it is the unification of a sense organ with its object, or a desire with its object. This is because unification requires the triad of the subject, the object and the process of unification. Unifications are eventually the different types of samadhi, from fleeting samadhis to sustained samadhis, from the chapter on Yoga. Whether the unification uplifts us or brings us down depends on the purpose it is fuelled by. We saw that as long as the triad of the unification is based on clean fuels, it forms an upward pointing triangle that can take us to a higher triad of stability, until we eventually reach the transcendental triad. From the analogy described earlier in the book, it is the establishment of an uplifting Prana that takes us to the balance or Sattva of a higher triad. In the following sections we will see some techniques used in Tantra to enable this across the various levels of life.

Higher and Lower Soma

The concept of the unification of a couple across various levels has caused some confusion with regards to the physical act of sex as a form of Tantric spiritual evolution. In order to see whether there is any evolutionary merit in a particular action, we have to first understand the concept of upward and downward pointing triangles. Any act which is done in excess of what is required for the purpose of evolution naturally forms the basis of a downward pointing triangle. It increases the power of manifestation as opposed to evolution. This principle applies equally to all acts, whether it is the act of eating or unifying with food, or the act of physical sex. Evolution requires Sattva across all levels of manifestation, and this includes physical and instinctual balance.

A branch of Tantra, known as 'left handed' Tantra (as opposed to the 'right handed' approach of this book generally), can use sex for spiritual purposes on rare occasions. However it requires many esoteric techniques as well as a very specific nature and karma, and so it is generally not relevant for today's society. Left handed Tantra makes use of sexual procedures as there is much power in it that can be harnessed. It contains the most fundamental power of the physical survival of a species, and it represents its greatest act of physical creation. Thus it can also be very negative when practised unnaturally.

The goal of life is about experiencing Soma or peace and happiness, and ultimately it is about the Soma or Ananda of Purusha. If we focus on experiencing Soma via the lower koshas such as via excessive physical consumption (dirty fuels), we create the basis of our reality there as we saw in the section on samadhi. We mistaken Soma to be physical, or external to the consciousness of samadhi within us, and so we are unable to directly experience it. Soma originates in the bliss (highest) kosha within us, and it is bliss of reality that ultimately underlies all manifestation. So any aspect of nature is able to offer it to us, and the direct experience of it fulfils our powerful and natural urge towards reality. Until we experience Soma in this way, we naturally cannot fully escape from the feeling of discontentment and depression. In addition to the meditation practices of Yoga, and the alchemy of Ayurveda, in this chapter we will see how Tantra can give access to Somas at the higher levels.

Dasha Mahavidya (The 10 Great Wisdom Goddesses)

In order to get a feeling for the techniques of Tantra, it is first useful to see how Shakti manifests itself in nature. Her energy is responsible not only for manifestation, but also for the diversity in manifestation. She unfolds this diversity according to the principles and intelligence in nature, such as the laws of dharma and karma. These laws allow for diversity by providing the framework for the many different harmonious rhythms of nature. Nature's rhythms are known as ritam in Sanskrit, and it is the perfect harmony of the laws that are evident in science, and it is the harmony in the rhythms of music. It underlies the inspiration in art, and it is the power behind mantras, especially the Vedic mantras, which are composed strictly according to various rhythms. The beautiful harmony of the ritam is ultimately the divine delight or Soma of Purusha. As such this dance of manifestation is considered to be like a play, which is known as leela in Sanskrit. It is symbolised by the image of Krishna (a form of Ishvara) playing the flute, and his music is filled with the ritam and Soma of Shakti. In a fiercer form that enforces change, the ritam is the Shakti of the dance of Shiva as nataraja, or Shiva as lord of dance.

This leela begins with Shakti taking on 10 different forms as two complementary pentuples. They are commonly known as the 10 great wisdom 'Goddesses' (though more accurately they are Devis), because they give us access to the knowledge required in order to understand the leela. The 10 forms each present a different aspect of this knowledge, and we have natural preferences for the different aspects according to our individual karma and dharma. A brief description of the 10 forms of Shakti is given below.

Krishna sounding the cosmic ritam on the flute.

Shiva as Nataraja, or lord of the cosmic dance.

1. Kali – Kali is the first of the wisdom Goddesses, and she represents the rhythms of time. All manifestation begins and ends in time, hence she comes first. As she is responsible for ending things, she rules over death and has a fierce nature.
2. Tara – The second wisdom Goddess is Tara, and her name literally means ray or wire. She represents the rays of Shakti's manifestation via thoughts and sounds, or via mantras. Physically this is represented by the rays of stellar light as we saw in the chapter on Astrology, and thus the word for star is Tara in Sanskrit.
3. Tripura Sundari – Her name means the beauty of the three worlds, and she symbolises the harmony or the perfect balance of the triad. She is the three-fold play of nature, and she is also the harmony or beauty hidden within it. Her knowledge of the triad reveals the beauty hidden in all levels of manifestation, and ultimately it reveals the beauty or Soma of the three-fold Purusha as Satchitananda.
4. Bhuvaneshvari – As Kali is responsible for time, so Bhuvaneshvari is responsible for the complementary concept of space. The wider definition of space is that of a mental field which contains all sensations, whether material or subtle. As space is the container of all, Bhuvaneshvari is referred to as the golden womb of all levels of manifestation.
5. Bhairavi – She is the highest form of energy or Agni, or the first form of energy of Purusha. She is Shakti in

her own essential form as energy, and as such a primal energy she relates to the energy of kundalini, as we will see in the next section.

6. Chinnamasta – The meaning of Chinnamasta is one without a head, and her energy is a fierce and electrical one. It destroys our sense of personality or the mental conception of our self in our heads. Instead she gives us the perception of consciousness beyond the mind and body.
7. Dhumavati – She is responsible for obscuring manifestation and giving us the vision of reality instead. She has the effect of putting the outer world into a state of deep sleep for us.
8. Bagalamukhi – Yoga tries to achieve the stillness of the unified state of samadhi, and it is Bagalamukhi that is responsible for bringing things to an instant standstill. She is the stunning power of reality that leaves everyone standing still in awe.
9. Matangi – Matangi bring the power of the unmanifest or causal word of Tara into manifestation, and so she is the power of mantra in action.
10. Kamalatmika – As Tripura Sundari symbolises transcendent beauty, Kamalatmika symbolises the manifest aspect of transcendent beauty. She is the Soma in life, including in everyday life.

The dashamahavidya represent an important means of guiding and channelling the energy of our evolutionary efforts according to our individual strengths and abilities. The wisdom Goddesses have many energetic techniques associated with them, and the practice of these helps increase their energy and knowledge within us. For on more on this, the book 'Tantric Yoga and the Wisdom Goddesses' by David Frawley contains many excellent insights into the dashamahavidya generally. It explains some of these techniques, and it also looks at the formation of pentuples and triads within the dashamahavidya.

Kundalini, the Primal Energy

Any evolutionary method or path of Yoga requires activity, or the use of energy. We saw that Agni as the cosmic fire was the main evolutionary guide. To this extent we could say that Agni is the metabolic or evolutionary aspect of Shakti, and that Shakti is the activity aspect of Agni. Thus manifestation as the field of evolution is the field of activity or Rajas guna. The path of Raja Yoga uses energy indirectly, as it aims at stillness and samadhi via mental techniques and implicit in this is an effort or a use of energy. By contrast Tantra focuses directly on energy itself, and the most primal form of energy or Shakti is symbolised by the Goddess Bhairavi. In us this type of energy is our very foundation or root, and therefore it is considered to be seated at the very root of our being or at our natural seat at the base of the spine. Shakti in this form is known as kundalini, and so Tantra is primarily about the management of the kundalini Shakti. It is via kundalini that Tantra attempts to unite the duality within us.

The duality within us is symbolised as the kundalini Shakti of Bhairavi at the base of the spine, and the bliss of perfect (Shiva) consciousness at the top of the head symbolised by Tripura Sundari. This dual is kept separated by the force of Prana, which as a result causes the manifestation of energy and life. The duality can be collapsed into a unity by the effort required to lift kundalini from the base of the spine to the top of the head, and this is the main goal of all Tantric techniques. When Bhairavi reaches Tripura, we achieve the knowledge of Kali (all time) contained within Tripura. This is the ability to transcend time to Shiva as Purusha. In terms of the Rig Vedic metaphors, Agni (at the base of the spine) melts Soma (at the top of the head), and the result is the flow of Soma as Vayu (in its form of pure Prana) to the state of transcendence.

The main effort required in raising the kundalini upwards lies in shifting the emphasis of our primal energy, or the primary purpose of our being, from the field of manifestation to the field of consciousness that underlies manifestation. If our primal energy is directed towards manifestation, it will naturally lie at the seat of our manifest being or at the base of the spine. From there we can use this energy to be involved in further manifestation via our activities, such as the act of reproduction which also has its centre around the base of the spine. Once we shift our primal focus towards reality and consciousness, our primal energy will naturally move towards its direction.

The shift in focus of our primal energy towards consciousness requires retracing the path of the projection of manifestation by consciousness. It is the evolution of the energy of the five elements, as opposed to their involution. It is the process of creating a series of upward pointing triangles based on clean fuels such as bija mantras, which we will see in the section on Mantras and Yantras. To this end we can trace the Shakti of the earth element back to the water element, the water element to the fire element, the fire element to the air element, and the air element to the space element. Each of these stages represents an expansion in conscious energy as the reversal of the waves of five-fold manifestation represented by the five elements. Finally we are left with the last triangle of the cosmic triad of Shiva (Purusha), Shakti (Prakriti as the sum of the five elements) and Prana. The collapse of this final triangle is the bliss of transcendence of the union of Shiva and Shakti.

The Chakras

The process of manifestation and evolution via the elements and Shakti is facilitated by the energy system of the chakras[4] (discs). The chakras are part of the subtle nervous (nadi) system mentioned in the chapter on Ayurveda, and they are areas where the nadis concentrate in a form similar to a disc. The nadi system represents the movement of the energy of conscious required for manifestation and evolution. Each of the five elements has a chakra associated with it, and the couple of Shiva and Shakti also have two further chakras associated with them. This represents the complete septuple from the dense earth element to the level of transcendence. Thus the raising of the kundalini is the movement of the kundalini up from the lowest chakra of the earth element to the highest chakra which symbolises pure Shiva consciousness. The next section looks at some energetic techniques associated with the chakras and the elements, and the movement of kundalini through them.

As the chakras are a means of the manifestation of the energy of consciousness, they are associated with the important organs and glands of the body. The chakra of the earth element is associated with the anus, the water element with the sexual organs, the fire element with solar plexus and the digestive fire, the air element with the heart, and the space element with faculty of speech or words. The chakra associated with Shakti is known as the third eye and is associated with the pituitary gland, and the chakra of Shiva is known as the crown chakra and is associated with pineal gland.

The goal of Hatha Yoga is also this raising of the kundalini to the crown chakra in order to facilitate samadhi consciousness. Its focus on physical asanas is simply to prepare the physiological aspects of the body associated with the chakras for the upliftment of the kundalini. Apart from asanas it also has many Pranic and energetic techniques for this purpose.[5] In Raja Yoga the raising of the kundalini is implicit in its meditation practices, as already mentioned.

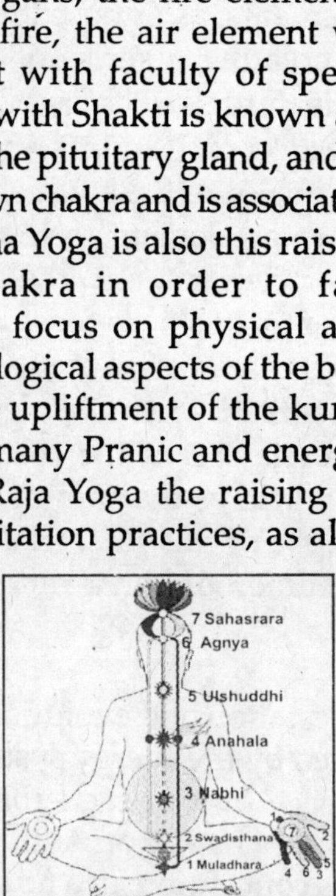

The seven chakras in relation to the human body, and their Sanskrit names.

Mantra and Yantra

In the chapter on Mantra we saw that mantras are an outpouring of the energy of Shakti. We saw further that they are the first such emanation of Shakti as they are associated with the first element of space, and therefore that they are the main instruments of causal activity. This makes mantras the prime tools in order to work with energy. Via the three levels of speech of pashyanti, madhyama and vaikhari, they connect the universal ritam of Shakti from the causal to the physical. The connecting ray or wire is symbolised by the Goddess Tara. In addition to speech, she also uses the rays of star (tara) light to facilitate this connection into the physical world. This is an important factor that underlies Vedic astrology, and it is also why mantras form the main therapy of astrology.

The main Shakti mantra is the bija mantra 'hreem', which we saw in the Mantra chapter. From the composition of its root sounds we saw that it was a very energetic mantra. It is the energetic equivalent of Om, the main transcendental mantra. One of the main uses of mantras in Tantric practices is that of the chakra mantras in order to help raise the kundalini. They form the basis of a clean fuel at the level of each of the chakras, and the resulting upward pointing triangle that is formed helps to uplift the kundalini. This is equivalent to expanding the awareness of consciousness within us. Each of the chakras of the five elements has a bija mantra associated with it, and these are the sounds of the semi-vowels as bija mantras. The semi-vowels are the linking sounds between the vowels which represent the highest koshas, and the consonants which represent the lowest koshas. The chakras similarly link the energy of the physical body to the energy of the causal levels.

The mantra associated with the earth chakra is 'lam' (all the chakra mantras rhyme with the English word hum). The earth element is the main form of support, and we saw the 'l' sound is also one of the most supportive sounds. The mantra associated with the water chakra is 'vam'. At the physical level it is water that carries the life-force or the Prana of the instincts. For example when we add water to the earth element, we see the first manifestation of life as plants. From earlier we noted that Prana is also related to Vayu, with Vayu symbolising the positive motion and activity that arises due to life. Vayu consists of the two root sounds 'va' and 'ya', and they relate to the pure and cooling motion of air the most of all the root sounds. 'Va' is the more supportive of the two by the configuration of the mouth, and thus it relates to life-force in the water element. It forms the basis of the mantra of its chakra 'vam'. Ya relates more directly to the air element, and forms the basis for it chakra mantra 'yam'. In between these chakras lies the chakra of the fire element, and it naturally relates to the most fiery of roots 'ra', which forms its mantra 'ram'. Finally there is the mantra of the space element, which is 'ham'. Ha is the root which expels the air from the body leaving only the space element in its place.

This establishes a rationale for the chakra mantras, and they are many different ways by which they can be used. To this extent we should note that raising the kundalini via the energetic approach is very difficult, and as with any advanced Yoga practice, it requires much work, discipline and understanding, as well as with appropriate guidance. Incorrect and rushed Tantric practices naturally lead to correspondingly dangerous imbalances of energy. However it is useful to gain a general familiarity and understanding of how these mantras work in the evolutionary process. This can lead to some gentle practices[6] which are generally not harmful. Initially it is possible to begin with a light and natural usage of these mantras for exploratory purposes, and this can be accompanied by a visualisation of the appropriate chakra element.

In addition to mantras another important technique of Tantra is the usage of yantras. They are geometrical patterns made up of upward or downward pointing triangles, often as an interplay between them. They are the light energy equivalent of mantras, and as mantras bring the ritam of Shakti from the causal world into the audible world, so yantras bring the causal ritam of Shakti into the world of physical imagery. Instead of infusing our minds with the evolutionary energy of the natural sound of a mantra, they enable us to infuse our minds with the evolutionary energy of the image of the yantra.

The primary yantra is known as the Shri Yantra or Shri Chakra. It is an interplay of a pentuple of upward pointing triangles, and four downward pointing triangles along with a dot (bindu) in the centre. The upward triangles represent the evolutionary force of Agni and Shakti across the pentuple of manifestation. The four downward triangles along with the bindu represent the cosmic pentuple (as noted in the chapter on Cosmology) coming into manifestation due to the force (Prana) of Shiva as Purusha. There is a bindu instead of a fifth downward triangle as it signifies the transcendent nature of Purusha beyond triads upon which manifestation rests. The bindu represents the seed of the cosmic triad, and along with the four downward triangles it forms the complete septuple. Thus the interplay of the triangles of the Shri Chakra naturally represents the cosmic ritam of the causal dance of Shiva and Shakti (known as the tandava), and the correct concentration on the Shri Chakra has the power to unfold this cosmic ritam with us.

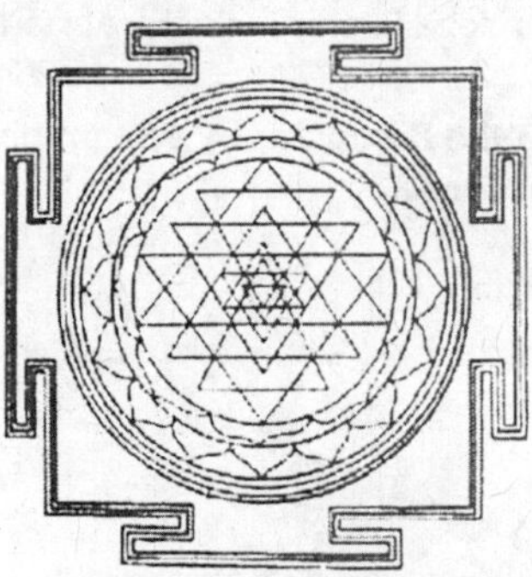

The Shri Chakra.

Vedic Alchemy

The concept of alchemy underlies all Vedic thought and practices. In Tantra it is the process of transforming a lower energy to a higher one, i.e. the instinctual energy at the base of the spine to the energy of pure consciousness at the top of the head. The poetical verses of the Rig Veda also refer to this process, and in Ayurveda the doshas are transformed to their higher functions via the practical methods of daily life. Yoga and Vedanta aim for transformation too, albeit by emphasising the mind before energy. All of these practices aim at transforming the individual into the universal, or seeking the real in the unreal, or experiencing the bliss of Soma in a world of decay, or becoming immortal in a world subject to time, or as it is often symbolised outwardly, by turning lead into gold.

In the next chapter we shall see another method of alchemy, which is the mystical ritual of inner sacrifice or Yajna.

ENDNOTES

1. Some examples of this are in Rig Veda 2.15 and 9.96.
2. Mahavidyasutram by Ganapati Muni is an example of this. At present it is only available in Sanskrit.
3. See Abhinavagupta by GT Deshpande.
4. Note Arthur Avalon's (John Woodruffe) books for information on chakras and kundalini.
5. Refer to the Hatha Yoga Pradipika of Swatmarama for more on this.
6. Note the books in the Bibliography section on Tantra that can be used for this purpose.

10.

Yajna: The Mystical Ritual of Inner Sacrifice

Adorable Agni, the Priest of the Sacrifice, the God of Ritam. The Bringer of Mysterious Treasures.

Rig Veda 1.1.1

Sacrifice at the Universal Level

Yajna means sacrifice, and the practice of Yajna is to look at life from the perspective of sacrifice. It is not sacrifice in the sense of suffering and hardship, but sacrifice as being a part of the universe's rhythm, and as a result sacrifice as a means of progress, fulfilment and contentment. We can see that all of nature's activity, its rhythms and its movements, are of the form of a sacrifice.[1] Activity is executed by the law of cause and effect, and more generally by the law of karma. Every cause must result in an effect, and every effect must in turn become a cause. So we see sacrifice at the heart of the universe's mechanism, as a cause sacrifices itself to its effect, and the effect to a subsequent cause. The law of karma is a sacrificial law, and this makes Karma Yoga the Yoga of self-sacrifice. Thus Yajna is not really about any outer activity or ritual, but an inner self-sacrifice. It is about harmonising with nature's law, it ritam, i.e. its general evolutionary purpose. As such Yajna is not about self-suffering, but about the bliss of being the Self.

As Yajna is at the heart of the law of karma and the universal evolutionary purpose, it is also at the heart of the way by which the individual once again regains the universal. It is the way by which the uplifting upward pointing triangles are formed, which make use of clean fuels for a pure flame of Agni and an uplifting and enlivening Prana. Yajna's goal of upward pointing triangles gives it a three-fold nature. Firstly there is the offering or performer of the Yajna as the fuel or Soma, then there is the execution of the sacrifice as the activity of Agni, and finally there is the resulting upliftment of Vayu to a higher level of Soma. Due to the role of Agni in effecting the activity of nature, it is also central to the process of Yajna, as we shall see.

The universal nature of Yajna means everything is always immersed in it. For example we see that the Sun receives energy as gaseous fuels, which it then offers as a sacrifice in the form of heat and light equally in all directions, including to us on earth. Once the energy of the Sun is received on earth, it is offered on through the various kingdoms of nature. The minerals pass on this energy to the plants, which in turn form the food of the animals and humans. In the humans the food enters the digestive tract from where it is offered to the blood, and from the blood it is offered to the other tissue elements of the body till it reaches the brain. From the brain it is offered to the mind via the action of the tanmatras. The various methods of Vedic alchemy then help us offer it on to the Devas within us, till we are perfectly evolved as Purusha. Thus Yajna is the actual route of samadhi and Soma of Yoga and Tantra. Similarly, every breath of life we take is also a sacrifice. The life-force in the breath offers itself to us, and through us it is offered on till it reaches the Devas, and finally it is offered back to Purusha.

If we break this chain of the universal sacrifice, i.e. if we consume what is offered to us without making an onward sacrifice, we attempt to break the inviolable law of karma. As rationality itself says this is impossible, we are forced by the laws of nature to make sacrifices against our wishes. The result is a downward pointing triangle of Vata and decay based on dirty fuels. As a result we suffer from pain, disease and loss, instead of being uplifted by the methods of Vedic alchemy. Therefore Yajna is the universally uplifting way for a subject to consume its object, whether it is our senses consuming the sense objects, or our mind consuming thoughts. True Yajna has the effect of uplifting the consumption of the object by the subject to the unification of samadhi.

Ashvamedha, the Horse Sacrifice

Examples of universal sacrifices are abundant in Vedic literature. One of the foremost of these is the Ashvamedha, or the horse sacrifice described in the first verse of the Brihadaranyaka Upanishad. The horse is often used as a metaphor for the potential of raw energy, or the life-force, as we saw in the chapter on Devas. In this verse the horse's head is described as the dawn, its eyes as the sun, its breath as the life-force, and its body as time. Thus this sacrifice symbolises the sacrifice of all the Prana in the universe to the only possible offeree of such an offering, i.e. Purusha beyond manifestation. Despite these fairly transparent metaphors, the horse sacrifice of the Brihadaranyaka Upanishad has often been confused with an actual killing of a horse in modern thinking. It is simply a metaphorical representation of the universal evolutionary process, or the universal dharma via Yajna.

Sacrifice at the Individual Level

We have seen that universal sacrifices affect us all, and now we shall see what sacrifice means for us from the perspective of the individual. The way we are able to harmonise with the universal sacrifice or the universal ritam, is via the concept of the individual sacrifice. As the universal sacrifice is the universal dharma, the individual sacrifice is the highest dharma at the individual level.

The individual sacrifice is also naturally based on the triad of Agni, Vayu and Soma. The main executor of the sacrifice is the activity of Agni; and here it symbolises our will or effort to uplift ourselves. Soma is the offering or the fuel, which in this case is ourselves. Any resultant progress or upliftment is symbolised by Vayu, which takes us to a higher level of Soma or contentment. Therefore if we have a clean purpose or will, we can make progress. The individual sacrifice begins even from such a simple and foundational level, and so any simple act can become a form of Yajna. It does not require any special act of outer self-sacrifice, though these may be of help in certain circumstances.

Karma Yoga is the main way of performing individual sacrifice. We sacrifice ourselves completely to the cause of upliftment with the aim of becoming conscious of the bliss kosha, or becoming conscious of the causal level where the real doers as the Devas exist in us. To this extent any action done with the causal level or reality in mind, rather than our emotions and outer personality, is a form of Yajna. At an outer level there are rituals that can help with this process symbolically, and these are the outer interpretations of the Vedic metaphors as we shall see in the section on Vedic Yoga. These outer rituals are not the main purpose of Vedic Yoga as has been emphasised, but they can serve as an entry point to it.

We see many examples of individual sacrifices portrayed in metaphoric form across Vedic literature.[2] There are stories were students offer their teachers, whether human or divine, cows, bulls and horses. This of course does not actually involve any animals, but it is a metaphorical representation of the qualities of the students themselves. The cow symbolises the offering of the clean mind as the fuel, the horse its activity, and the bull its power. These are the qualities that produce the pure and uplifting flame of Agni. Hence these stories are the individual equivalent of the universal horse sacrifice.

Yajna and Vedic Yoga

The concept of Yajna is one of the main images set forth in

the Rig Veda, and this image also occurs across Vedic literature generally. The Vedic text that specifically deals with Yajna is the Yajur Veda, which is considered the second Veda after the Rig Veda, with all its verses relating to Yajna. The third primary Vedic text is the Sama Veda, which is a presentation of the Rig Vedic verses in a more melodious form. Together these three Vedic texts form the triad necessary for Yajna, and they form the basis of the Vedic Yoga, which is a means of performing the inner individual sacrifice. The performer of the sacrifice is the offering of the lower Soma or fuel, the mantras of the Rig Veda is the Agni that exercises our will, the mystical sacrificial rites of the Yajur Veda are the acts required for upliftment, and the melody of the Sama Veda is the resultant higher Soma. Therefore these three Vedas are not really separate, but form a complementary whole as a process of Yoga via an inner sacrifice to the level of the Devas within us.

The metaphorical nature of the Vedic verses, as well this mystical method of performing a Yajna, reflect the multi-dimensional nature of the Vedic texts. It indicates how the Vedic verses can act as seeds for the more practical aspects of the Vedic system. It underlies the practices of Raja Yoga and Tantra, and it is the thinking behind Vedanta. Via Agni, Vayu and Soma, and along with the many astronomical references relating to the Devas, it also underlies Ayurveda and Vedic Astrology.

Even if we follow the outer aspect of the Yajnas described in the Vedas, they have some evolutionary or dharmic merit. The outer Yajnas involve making a physical fire with wood and ghee as the fuel. In this way the earth and water elements are converted to the energy element. Due to the special way of using the wood, ghee and certain herbs as the fuel, the energy element produces a clean vapour that purifies the space element. Thus we can see that the evolutionary process is assisted via the evolution of the elements, and so we can see that such an outer Yajna can also be dharmic. For this reason it is performed to help facilitate the smooth fulfilment of karma during important or difficult moments, such as births, deaths, marriages, diseases, spiritual development etc. However we should note that it is simply an outer act, and it is only the entry into the deeper Vedic tradition. It is the inner Yajna where the real power lies, and which underlies the various Vedic systems.

The meaning and level of the Vedic Yoga of Yajna has for us will depend on the level of our understanding of reality. We saw that if our samadhis are mainly at the physical level, then our understanding of reality is also sensory and physical. Yajnas for us will then also be mainly at an outer level, i.e. ones that are based around a physical fire. If we tend to have samadhis at the level of the intelligence kosha, then Yajnas for us will involve our inner revelatory fire. If our samadhis are rationally dominated, Yajnas for us will involve the fire of logic and pragmatism. Our Yajnas simply reflect the level we feel religious at.

Yajna, Purification and the Environment

Generally our sense of the concepts of clean dirty is based upon our purpose in life. Since our primary purpose is to remain alive with healthy physical bodies, our sense of cleanliness is centred around that, i.e. cleanliness as physical hygiene. Ultimately our deepest and most powerful purpose is our sense of progress and evolution, or our karma and dharma. Thus our deepest definition of cleanliness will be one that reflects this purpose, and so it will include hygiene at a mental level too. To this extent inner Yajnas naturally also act as a form of mental purification. We can see a good example of an inner Yajna as purification in Patanjali's Yoga. We saw that the second of the eight limbs are the five niyamas or set of mental observances. These were: 1. Self Surrender 2. Self Study 3. Use of the Will or Tapasya 4. Purification 5. Contentment

Self surrender is like the offering of a sacrifice or the lower Soma as ourselves. Self study and use of the will are ways of burning the inner Agni. They are the effort required for positive Agni once we have surrendered ourselves to its purpose. The purification is the upliftment of Vayu, and the final niyama of contentment is the higher level of Soma we experience as a result of the sacrifice. So we can see how an inner Yajna fulfils our deepest definition of purity via the practice of Yoga.

An outer Yajna must also have an outer purifying element to it, as it has a sense of evolution via the progression through the five elements from earth to space. To this extent it tends to purify the space and atmosphere around where it is performed. It can help in cases where an inner Yajna cannot be performed, especially where the mental control has been reduced. However an outer Yajna may not always suit the evolutionary purpose. For example if we have a greatly reduced number of trees and a shortage of wood as a consequence, the fuel required for it will cause further environmental imbalance. It will cause the environment to be unsuitable for sustaining life, and it will be detrimental to evolution. It is possible for another fuel to become suitable instead, but we do have to keep our sense of clean fuels updated according to the times we live in, especially in the outer world where conditions can change more quickly than in the inner world.

As we generally do not see our deepest purposes easily, we are not always good at judging what is clean for us at the deepest level, or the decay and disease that poor hygiene at this level can cause. For example we still do not fully recognise that over-consumption is an unsuitable purpose for us. We do not see that the evolutionary purpose of our actions and our consumption is to lead to our higher selves (the Devas), and thus we have no sense of sacrifice for this purpose. Once the level of sacrifice is not in balance with the level of consumption, the smooth fulfilment of the law of karma is no longer possible. It is effectively the opposite of a Yajna, or a downward pointing triangle of Vata and disease, instead of the upward pointing triangle of Vayu and life giving Prana. It results in the forced fulfilment of karma which can cause disasters at an individual and global level.

These can be avoided if our actions suit our karma and dharma. We must consume only what we need to evolve mentally, which is the positive concept of sacrifice as one that uplifts us, and one that leaves us feeling contented. It seems that traditional societies had this knowledge and insight judging by their approach to consumption, and their respect for the environment and nature generally. As the Vedic tradition focuses on Yajna and its evolutionary power of life, it is no wonder that in India it continues to have the strength to underlie the oldest existing civilisation in the world today, where its ancient practices can still be seen.

ENDNOTES

1. Chandogya Upanishad 3.16-17
2. An example is in the Brihadaranyaka Upanishad 3.1.1-2.

Afterword

The main ideas behind this book can be summarised in just a few words:

- Ultimately knowledge is only directly perceivable, and not deducible. (Chit)
- This is the realisation that the individual (Pindanda) and the universal (Brahmanda) are identical. (Sat)
- It is the completeness of reality, and it is harmony, peace and happiness. (Ananda)

Whether it is the verses of the Vedas, the philosophy of Vedanta, the samadhi of Yoga, the Soma of Tantra or the offerings of a Yajna, they all aim to reveal this to us.

Nature itself reveals this philosophy to us when we recognise ourselves as being nature, rather than as being owners of nature. Then we are able to experience the role of the Sun, the Moon, the wind, the rain and the great elements, not just as forces outside us, but as integral parts of our inner nature. We not only experience the scope of nature's laws such as rationality, but we also experience its deepest purpose, which is our purpose. This is the inspiration behind the Vedic verses of the Rishis and their Devas.

The following verses are a Shanti (peace) Mantra in Sanskrit that express this purpose for all of us:

Om sarvesham svastir bhavatu
sarvesham shantir bhavatu
sarvesham purnam bhavatu
sarvesham mangalam bhavatu

Om Harmony be to all,
Peace be to all,
Completeness be to all,
Auspiciousness be to all.

sarve bhavantu sukhinah
sarve santu niramayah
sarve bhadrani pashyantu
ma kashchit dukhbhagbhavet

Let all be harmonious,
Let all be healthy,
Let all see goodness,
Let there be no difficulties and sadness.

Om Shantih Shantih Shantih

Bibliography and References

Classical Works (By Title)

Ashtanga Hridaya of Vagbhatta, *Translation of KR Srikantha Murthy*

Bhagavad Gita (God Talks With Arjuna), *Commentary of Yogananda Paramahansa*

Bhavaprakasham of Bhavamishra, *Translation of KR Srikantha Murthy*

Brahma Sutras of Badarayana, *Commentary of Swami Shivananda*

Brihadaranyaka Upanishad, *Translation of Swami Madhavananda*

Brihat Jataka of Varahamihira, *Translation of Suryanarian Rao*

Chandogya Upanishad, *Translation of Swami Gambhirananda*

Charaka Samhita of Charaka, *Translation of PV Sharma*

Hatha Yoga Pradipika of Swatmarama, *Translation of Swami Muktibodhananda*

Paratrishika Vivarana of Abhinavagupta, *Translation of Jaidev Singh*

Rig Veda Samhita, *Edited by RL Kashyap*

Sankhya Karika of Ishvara Krishna, *Translation of Swami Virupakshananda*

Saundarya Lahiri of Shankaracharya, *Translation of Swami Tapasyananda*

Shiva Sutras, *Translation of Jaidev Singh*

Surya Siddhanta, *Translation of Danavir Goswami*

Tripura Rahasya, *Translation of Swami Ramananda Saraswathi*

Upanishads (The Principal), *Translation of Sarvapalli Radhakrishnan*

Vakyapadiyam of Sri Bhartrihari, *Translation of Satyakam Varma*

Vedanga Jyotisha, *Translation of Suresh Chandra Mishra*
Yoga Sutras of Patanjali, *Commentary of Swami Vivekananda*
Yoga Vasishta (Laghu) of Abhinanda Pandita, *Translation of KN Subramanian*

Later Works (By Author)

Avalon, Arthur (John Woodruffe), Serpent Power
Avalon, Arthur (John Woodruffe), Shiva and Shakti
Charak, KS, Elements of Vedic Astrology 1 & 2
Deshpande, GT, Abhinavagupta
Frawley, David, Gods, Sages and Kings
Frawley, David, Mantra Yoga and Primal Sound
Frawley, David, Tantric Yoga and the Wisdom Goddesses
Frawley, David, The Astrology of the Seers, David Frawley
Frawley, David, The Wisdom of the Ancient Seers
Frawley, David, Yoga and Ayurveda
Goswami, Amit, Self Aware Universe
Kak, Subhash, Prajna Sutra
Penrose, Roger, The Emperor's New Mind
Sri Aurobindo, Hymns to the Mystic Fire
Sri Aurobindo, Secret of the Veda
Swami Rama Tirtha, In Woods of God Realisation
Swami Sada Shiva Tirtha, The Ayurveda Encyclopaedia
Swami Shri Yukteshwar, The Holy Science
Swarup, Ram, Meditations/Yogas, Gods, Religions
Swarup, Ram, The Word as Revelation
Wood, Ananda, From the Upanishads
Wood, Ananda, Ways to Truth: A View of Hindu Tradition

Glossary

Advaita: non-duality, undivided
Agni: universal principle of energy and change symbolised by fire
Aham: I
Ahamkara: I maker
Ahimsa: non-violence, non-harming
Ananda: real (transcendental) delight
Atman: individualised reality (Brahman)
Avidya: ignorance at the level of the individual
Bhakta: devotion
Bija: seed
Bindu: dot or point as a source
Buddhi: individualised intelligence
Brahman: reality, i.e. transcendence
Chakra: disc, commonly used term for one of seven subtle energy centres
Chikitsa: treatment or therapy
Chit: real (transcendental) consciousness
Chitta: field of manifest consciousness
Dharma: nature's rhythm or law
Dhyana: meditation
Dosha: weakness, one of three biological humours
Dvaita: duality, divided
Guna: fundamental quality
Jivatman: individualised reality as the reincarnating entity
Jnana: wisdom
Kapha: biological humour symbolised by water
Karma: actions and their results
Kosha: covering, especially relating to the various layers of manifestation
Kundalini: primal life energy normally at the base of the spine
Leela: cosmic rhythm as a delightful play
Madhyama: middle or mental sound
Mahat: universal intelligence

Manas: emotional mind
Maya: illusion of creation
Nakshatra: one of 27 stellar constellations normally relating to the Moon
Pashyanti: causal or creative sound
Pitta: biological humour symbolised by fire
Prakasha: transcendental equivalent of light
Prakriti: basis of all manifestation and nature
Prana: transcendental and manifest life-force
Pranayama: control of Prana
Pratyaksha: direct perception
Purana: ancient encyclopaedic texts
Purusha: real (transcendental) person, commonly a synonym for Brahman
Rajas: universal quality of activity
Rasa: taste or mood across various levels, subtle and physical
Rishi: 'seers' of wisdom that connect the divine and the human
Ritam: universal rhythm
Sakshi: transcendental witness
Samadhi: consciousness where subject and object are unified
Samskara: causal seeds
Sanatana Dharma: the eternal law or rhythm underlying Hinduism
Sat: reality or what is true
Sattva: universal quality of balance or harmony
Shad Darshana: six philosophical view points of manifestation
Shakti: transcendental and manifest energy
Shanti: absolute peace
Soma: universal principle of nourishment and delight symbolised by water
Tamas: universal quality of inertia
Tattva: principle
Tanmatra: mental or inner aspect of the senses
Upanishad: the final philosophical section of the Vedas
Vaikhari: varied or audible sound
Vairagya: detachment from the unreal
Vata: biological humour symbolised by air
Vayu: universal principle of motion symbolised by air
Viveka: ability to discriminate between the real and the unreal
Yantra: instrument, a geometrical pattern that aids meditation and realisation
Yuga: epoch or an age

Index

I

J

K

L

M

N

O

P

Q

R

S

T

U

V

Y

BOOKS OF RELATED INTEREST FROM NAB

THE ESSENTIAL VEDĀNTA : A New Source Book of Advaita Vedānta—*Eliot Deutsch, Rohit Dalvi (eds.)*

Here is a resource for everyone interested in the study of Indian Philosophy, containing valuable collection of Advaita Philosophic literature.

This book is a comprehensive introduction to the tradition of Advaita Vedānta based upon primary sources and makes possible a study of it in its classical form. Translations from all major Sanskrit writings are included here along with the selections from all of the most important representatives of this central philosophy of India.

ISBN: 978-81-7822-265-3

THE PRINCIPAL UPANISHADS : A Poetic Transcreation —*Alan Jacobs*

The origin of the Upanishads, sacred writings of Hinduism, predate recorded history. They are thought of by many as perhaps the greatest of all the books in the history of world religions. Alan Jacobs has used free modern verse to convey the essential meaning and poetry of the original text, omitting Sanskrit words as much as possible and providing a clear, contemporary commentary. He brings ancient texts to life, passing on the essence of this perennial message in a totally fresh way.

ISBN: 978-81-7822-238-7

VEDIC EQUALITY & HINDUISM : A Reformist Agenda: Dalit Emancipation and Return to Vedic Brotherhood —*O.P. Gupta*

In this book, the author has given expressions to his understanding of the original and genuine caste system, the Vedic way and approach to human life, the tradition of Sati, its misunderstanding and misrepresentation and wrong notions spread by *Manusmriti*. According to him, the original teachings of Vedas are that all Hindus are equal by birth i.e., there is complete gender as well as caste equality. He feels concerned about the Dalits and their emancipation as also about the ills that beset the Hindu society today and is deeply committed to their removal. A must read for all those in search of their roots.

ISBN: 978-81-7822-285-1